The Sugar Trap

Books by Beatrice Trum Hunter

The Natural Foods Cookbook

Gardening Without Poisons

*Consumer Beware! Your Food
and What's Been Done to It*

*The Natural Foods Primer,
Help for the Bewildered Beginner*

Whole-grain Baking Sampler

Food Additives and Your Health

Yogurt, Kefir, and Other Milk Cultures

*Fermented Foods and Beverages,
an Old Tradition*

Favorite Natural Foods

*The Mirage of Safety, Food Additives
and Federal Policy*

The Great Nutrition Robbery

How Safe Is Food in Your Kitchen?

The Sugar Trap and How to Avoid It

BEATRICE TRUM HUNTER

The Sugar Trap

and How to Avoid It

Boston

HOUGHTON MIFFLIN COMPANY

1982

Library of Congress Cataloging in Publication Data

Hunter, Beatrice Trum.
    The sugar trap and how to avoid it.
    Bibliography: p.
    Includes index.
    1. Sugar.   2. Sugar substitutes. I. Title.
TX553.S8H86   613.2′8   81–20110
ISBN 0–395–31824–6   AACR2

Printed in the United States of America

s 10 9 8 7 6 5 4 3 2 1

To Ohana,
for new dimensions

Contents

The Sugar Trap

1

The Sugar Trap:
How Did We Get into It?

AVOID TOO MUCH SUGAR . . . Estimates indicate that Americans use on the average more than 130 pounds of sugars and sweeteners a year.
— "Nutrition and Your Health; Dietary Guidelines for Americans," U.S. Department of Agriculture / U.S. Department of Health, Education, and Welfare, February 1980

It is important to remember that refined and processed sugars have been added to a wide range of products. Although labeling regulations do not currently require the content of the different sugars to be described, if some kind of sugar (corn syrup, fructose sugar, dextrose, honey, etc.) is listed as one of the first two or three ingredients, then one can reasonably assume that there is a lot of sugar added to the product.
— *Dietary Goals for the United States,* prepared by the staff of the Select Committee on Nutrition and Human Needs, U.S. Senate, 2nd ed., rev., December 1977

ALTHOUGH WE ARE TOOTHLESS AT BIRTH, already we have acquired a sweet tooth. We have developed taste buds in our fourth month in the womb. Almost immediately after birth we respond to sweet tastes with smiles and to bitter tastes with grimaces. As newborns, we display the same positive responses to sweet tastes as we do later as adults.[1] With surprising precision, as infants we are able to distinguish different levels of sweetness, and like adults, we prefer high concentrations.[2]

An inborn preference for sweets is believed to be an evolutionary adaptive mechanism guiding us to nutritious fruits and vegetables that are high in energy-rich carbohydrates. Many other animal species also have learned the survival value in selecting sweet plants. Our aversion to bitterness in infancy is thought to be another survival mechanism, steering us away from toxic alkaloids present in many bitter plants.[3]

In recent times, however, a taste for sugar has become maladaptive. As we learned to extract and concentrate the sweetening components from plants, we succeeded in separating sweetness from nutrition. Our greatly increased consumption of concentrated sugars divorced from their nutrients is now incriminated in a wide range of health problems.

The sugar trap is baited early. At birth, we are offered sweetness, either with lactose, the breastmilk sugar, or with cane, beet, or corn sugars in infant feeding formulas. Sugars and sweetened foods continue to be offered in commercially prepared solid baby foods and toddler foods.[4] Our sugar consumption is greatest between our twelfth and fourteenth years. As we mature, generally we continue to enjoy sweet tastes, but culturally we acquire tastes for bitterness and for subtle mixtures of sweetness, bitterness, saltiness, and sourness.[5]

Indisputably, we humans enjoy sweetness. In some circumstances, it is believed that our enjoyment of the food's taste may affect how well we utilize the food. In this respect, sweetness may play an important role.

In our cultural traditions, sweetness has been associated with goodness. More than 60 English phrases include sugar, and as many include honey and other syrups, all conveying positive ideas about taste, smell, appearance, acts, and characteristics. Terms of endearment include "you're my sugar," "sweetie pie," "sweetheart," and "honey."

Sugars and starches are classified as carbohydrates and, along with proteins, fats, minerals, vitamins, and water, constitute our total food and beverage consumption. Nutritionists generally agree that a diet comprised of about 50 percent carbohydrates is normal and healthy. But today about one third of our total carbohydrate consumption is straight sugar. The sugar interests and their supporters regard this intake of sugar as "moderate." Since sugar is the cheapest and most plentiful food energy source available, proponents suggest that we could consume at even higher levels than at present without overstepping the bounds of moderation. They argue that, in a well-balanced diet, having enough calories provided by carbohydrates and fats keeps us from having to burn our protein for energy. Furthermore, they claim that sugar is eaten with foods and beverages, many of which contain specific nutrients, so the total dietary intake is not "empty calories." [6]

These arguments are specious. While both sugars and starches are carbohydrates, we utilize them differently. Sugars are simple carbohydrates while starches are complex. We metabolize sugars rapidly and starches slowly. The candy bar is noted for its "quick energy," while a baked potato provides slow, sustained energy. Total blood fat levels of triglycerides and cholesterol, considered as heart disease risk factors, are increased significantly in sugar diets but not in starch diets.

Is sugar a cheap calorie source, as sugar proponents would have us believe? While the growing of sugar may require low energy use, its refining is energy-intensive. According to the Census of Manufacturers' standard industrial classifications, the processing of sugar beets and the refining of cane sugar, along with the milling of wet corn and the processing of malt beverages — all sources of "empty" calories — accounts for 20 percent of the total energy used to process food in the United States. The production of one pound of refined beet sugar requires 4360 kilocalories, and cane sugar, 2610 kilocalories. Yet each returns only 1746 empty nutritional kilocalories. In Brazil, it was found that less energy was needed to ferment and distill sugar into alcohol than to refine it. [7]

Can we really view our present sugar consumption as moderate? Not by standards of human experience through many centuries, nor by those who still live on traditional diets. Prehistoric humans consumed carbohydrates mainly from wild fruits, berries, starchy tubers, and other plants. Infrequently, they found concentrated sugars from such sources as wild honey or licorice root, but such finds were so rare that they were recorded as special events in stone-age drawings and other artifacts. Currently, in scattered areas of the world where people still cling to their traditional foods, simple carbohydrate consumption is exceedingly low and infrequent.

In the human experience, carbohydrates have been consumed commonly as complex carbohydrate. Two major events reshaped this traditional pattern. The first was the technological development of sugar cane refining. The second was the development of the sugar beet during the Napoleonic Wars, as a solution to the British blockade of continental ports. The outcome of these two events was to shift us away from predominantly complex carbohydrates and direct us toward simple and refined ones.

Sugar proponents insist that our sugar consumption has not increased over the last hundred years but has remained at a level of more or less a hundred pounds per person annually, except during World War II when rationing forced consumption down. Our sugar consumption has *not* remained stationary over the last hundred years. This canard is founded on figures drawn from a base period of U.S. sugar consumption from 1910 to 1913, the time when collection of sugar statistics began. *Prior to that period, consumption was significantly lower.* As consumption began to increase dramatically, it became apparent that information should be gathered.[8]

As Americans, we are eating more and more sugar, eating more in refined form, and consuming much of it in ways that are not immediately obvious to us. The total sugar content of our diet, from all sources including naturally occurring sugar found in many foods, as well as that contained in syrups, honey, and from beets, cane, and corn, rose about 25 percent from the early 1900s to the early 1970s.[9]

According to a United States Department of Agriculture study, the increased use of sucrose (sugar) is largely traceable "to the desire of food manufacturers to create unique food products with a competitive edge." Among examples cited was the practice, begun in 1948, of adding sugar to breakfast cereals in order to boost slumping cereal sales. The effort was successful, and ever since "the profusion of varieties of cereals, soft drinks, and other products represent efforts to protect market shares." [10]

Not only are traditionally unsweetened foods now sweetened, but sweetness levels are much higher. For instance, formerly bread recipes did not have sweeteners among their ingredients, and early cake recipes had far lower levels of sweetening than current ones.[11]

A study by scientists from the Agricultural Research Service, USDA, released in 1974, showed that per capita use of refined sugars rose 33 percent since the beginning of the century. By the early 1970s, Americans were consuming about 102 pounds of refined sugars per person annually. By 1977, the total of all sugars and sweeteners, exclusive of non-nutritive ones, reached the staggering figure of 137.8 pounds per person annually.[12] Since then, the annual consumption has declined slightly.[13]

Statistics for sugar consumption are divided between its industrial use to manufacture processed foods and beverages and its home use. In 1925, almost two thirds of all sugar was used by consumers directly within the home, while only one third was used by industry. Currently, industrial use accounts for about 70 percent of all sugars used, while direct household use has declined correspondingly. What caused this shift?

Home decline was not from lower sugar consumption but rather from increased purchases of convenience foods. The manufacture of pre-sugared items and ready-to-eat products proliferated, especially since the 1960s. At present, the soft drink industry uses about half of all industrial sugar supplies, and the remainder is used mainly by bakers, canners, confectioners, and ice cream and dairy processors. As home sugar purchases continue to decrease, *we have less control over the levels of our sugar consumption than people did at the turn of the century.*

The 1973 USDA study also showed that higher incomes resulted in increased purchases of convenience foods. Additional factors contributed to increased refined sugar consumption. One was the higher proportion of teenagers and sub-teenagers in our population, a group likely to consume above-average quantities of sugar by as much as 20 percent. This means that, on average, they are consuming 140 to 150 pounds of sugar yearly. Other factors included the development of new uses for sugar. And the 1970 cyclamates' ban led to greater use of sugar in partial replacement of cyclamates.

Even using the 1910–13 base period cited by the sugar proponents, the radical transformation of our total carbohydrate consumption is apparent. During the 1910–13 period, the average American ate 498 grams of total carbohydrates daily, consisting of 342 grams from starches and only 156 grams from sugars. By 1974, the average American ate more fats and protein at the expense of somewhat fewer total carbohydrates. But the decline of total carbohydrates to 375 grams is less significant than the composition of the carbohydrates: Starch consumption declined to 197 grams while sugar consumption rose to 200 grams. In simple terms, we were eating fewer complex carbohydrates, such as potatoes and bread, and more simple carbohydrates such as sugar. Sugar's percentage of our total carbohydrate consumption in the 1910–13 period was 31.5 percent; by 1974, it had risen to 52.6 percent.[13] Thus, even using the sugar proponent's statistical base period, the rise in sugar consumption is well defined. A release by the Society for Nutrition Education in 1975 reported that sugar and high-sugar products had reached the stage that "the average American now consumes his or her own weight in sugar yearly."[14]

Another unjustified claim of the sugar defenders is that sugar is eaten with foods and beverages, many of which contain specific nutrients and make a contribution to the diet. In examining the types of foods and beverages that are the heaviest in sugar, we see that they are ones that lack other nutrients, too. Soft drinks, candies, and chewing gums account for one fourth of all sugar uses. Other products, such as frozen desserts, pies, puddings, and cakes, not only have high sugar levels, but also contain objectionably high levels of fats and/or refined flours, with insignificant levels of desirable nutrients, such as proteins, minerals, and vitamins.[15]

Sugar consumption is likely to increase due to our present dietary patterns, which, in turn, are altering radically some of our traditional cultural patterns. Home food preparation from basic commodities continues to decline as more women join the work force. Advertising has convinced many individuals that convenience foods save time and money, claims that are not always true. With more discretionary income, many people have turned to built-in convenience. At present, the major portion of the foods we eat is processed outside the home, and more than 70 percent of these processed foods contain added sugar.[16] According to predictions, continued increased consumption of such foods will continue and may be nearly total by the end of the century. Also, Americans are eating away from home more frequently. At present, one out of every three meals is eaten in a restaurant or other feeding institution. This trend, too, may accelerate, with the expectation that by the end of the 1980s, two out of every three meals will be eaten away from home. Cultural patterns are disrupted, as many Americans no longer sit down together as a family to share meals in common, but snack individually and erratically throughout the day. While snacks may consist of nutritious foods, many of those commonly being consumed are high in sugar. One junior high school teacher had his students analyze their families' spending habits, and they found that 20 to 25 percent of the food budget went for sodas, cookies, chips, candies, and cakes.[17] This pattern may be typical for a large segment of the population.

Sugars are well recognized in certain types of foods, where they are expected, for example, in cakes, candies, puddings, jellies, and frozen desserts. But sugars are not apt to be recognized readily in foods that traditionally did not contain sweeteners. Today, hidden sugars may be present in a wide variety of convenience foods purchased for home use or used in restaurant preparations, including cured ham, luncheon meats, and frankfurters; bouillon cubes, soups, and gravies; peanut butter; potato chips; dry roasted nuts; coatings for fried or baked poultry; restructured poultry; mixes intended to stretch the protein content of chopped beef; certain canned and frozen vegetables; frozen and canned entrees; plain cottage cheese; instant coffee, both regular and decaffeinated, and instant tea mixes; and roasted coffee. Even iodized salt contains some sugar to stabilize the potassium iodide. In tests, some catsups contained a greater percentage of sugars than did ice cream; some salad dressings had three times more sugar than did cola drinks; and at least one non-dairy creamer contained a greater percentage of sugar than did a typical chocolate candy bar.

New developments in food technology continue to contribute additional sugars to the American diet. For example, a frozen milk concentrate is being developed that, in the near future, may be marketed like frozen fruit juice concentrates. Most of the milk sugar (lactose) is removed from the milk concentrate and replaced with soluble sugars such as corn sweeteners or sucrose.[18]

Sugar's presence need not be listed on all food labels. Even when sugar is listed, consumers are unable to judge the amount present since percentages of sugar are not required to be listed. Nor do present nutrition labeling regulations require that the sugar portion be stated in a listing of "carbohydrates." For example, a fruit-flavored yogurt may list carbohydrates as 49 grams, compared to 17 grams for plain yogurt. Consumers cannot know how much of the 49 grams is derived from fruit and how much from the sugar in the fruit jam that flavors the yogurt.[19]

As more consumers learned that ingredients are listed on labels in a descending order of predominance, many were shocked to learn that some pre-sweetened breakfast cereals contained more sugars than grains, since sugar was listed first, even before the grain. The cereal processors took note. By reformulating and using a number of different sweeteners, they could maintain the same high sweetening level in the products, but the label information made it appear that grain, listed first, predominated. For example, the label of one hypothetical product might read: "Contains sugar, wheat, vegetable oil" (followed by other ingredients); after reformulation, the label might read: "Contains wheat, sucrose, vegetable oil, brown sugar, corn syrup solids, malt powder" (followed by other ingredients). The total sugar content may still be greater than the total grain content.

Consumers who read food and beverage labels carefully do not always recognize the presence of sugars that appear under many guises. All the following are sugars: sucrose, dextrose, corn syrups, crystalline fructose, high-fructose corn syrups, sorbitol, xylitol, mannitol, turbinado, raw sugar, brown sugar, molasses, sorghum, honey, maple syrup, malt syrup, maltol, and more. Some, such as maltodextrins, are unfamiliar. At times, misleading phrases are used, such as "fruit sugar" or "natural carbohydrates." While the *-ose* ending of words indicates that the substance is a sugar, not all sugars end in these letters.

Consumers remain ignorant about sources of hidden sugars that are used in certain industrial practices before foods reach the home or restaurant. Sugar is used by the meat industry to feed animals prior to slaughter, a practice reported to improve meat flavor and the color of cured meat prepared from pigs.[20] A dehydrated molasses blended with corn syrup may be added to hamburgers and other ground meats used in restaurants to help reduce shrinkage and improve the meat's flavor, juiciness, and texture. Breading, intended to coat seafood before it is deep fried, may contain sugar. Raw potato slices are dipped into sugar solutions before they are deep fried. Whole salmon may be glazed with a sugar solution before it is vacuum packed and frozen. Honey solutions may be injected into pieces of chicken before they are fried.

By the 1970s, increasing numbers of nutritionists, physicians, dentists, researchers, and public health officials were voicing concerns about the public health effects resulting from high levels of sugar consumption. A growing body of incriminating evidence associated high sugar intake with dental caries and obesity. Additional data suggested links between high sugar consumption and a wide range of other health problems.[21]

In 1973, the Select Committee on Nutrition and Human Needs of the Senate (commonly called the McGovern Hearings after its chairman, Senator George McGovern, D.–S.D.) held hearings on sugar in the diet. Professionals from here and abroad attested to the various health problems created by high sugar consumption, including such conditions as diabetes, coronary heart disease, hypoglycemia, and behavioral problems.

Information collected in these hearings, as well as in hearings held by the committee concerning other aspects of the American diet, led to the issuance in February 1977 of the *Dietary Goals for the United States*. Among the recommendations, Americans were urged to increase total carbohydrate consumption so that it would comprise 55 to 60 percent of the total caloric intake, but *to reduce sugar consumption by about 40 percent so that it would account for only about 15 percent of the total caloric intake*. In order to achieve these changes, the committee recommended that Americans increase consumption of fruits, vegetables, and whole grains and decrease consumption of sugars and foods high in sugar content.

The recommendations were reasonable and similar to dietary goals previously recommended officially by other countries.[22] Nonetheless, publication of the *Dietary Goals* sparked a sensational controversy that raged for months and continues to flare up spasmodically. Affected interests, such as the sugar industry and food and beverage processors, entered the fray. Some nutritionists also joined battle for various reasons, but mainly in objection to recommendations concerning fats and salt, rather than sugar. As a result, ten months later, the committee was forced to issue a revised version of the *Dietary Goals*. Although certain recommendations, notably concerning salt, were weakened, sugar reduction recommendations were strengthened. The committee urged Americans to increase consumption of complex carbohydrates and naturally occurring sugars from about 28 percent to about 48 percent of total caloric intake, but *to reduce consumption of refined and other processed sugar by about 45 percent to account for about 10 percent of total caloric intake.*

Ironically, these recommendations had a wry twist. Between 1970 and 1980, Americans became weight-conscious. Although they succeeded in reducing total caloric intake by about 10 percent, they achieved it by consuming a greater proportion of calories from sugar! The increased use of sweeteners, chiefly in soft drinks, actually increased per capita sugar consumption. This increase, combined with a lower total food intake, resulted in a substantial rise in the proportion of calories from sweeteners. Despite the drop in caloric intake, the incidence of obesity increased in our population.

Refined Sweeteners:
Sweet and Detrimental?

Cane, beet, and corn sugars

The refining of sugar may yet prove to have been a greater tragedy for civilized man than the discovery of tobacco.
—J.A.S. Dickson, M.D., *The Lancet,* August 15, 1964

If only a small fraction of what is already known about the effects of sugar were to be revealed about any other material used as a food additive, that material would be promptly banned.
—John Yudkin, M.D., *Sweet and Dangerous,* 1972

Sucrose: cane and beet sugars

PLANTS CONTAIN many types of sugars. You can identify them as sugars since the names use the suffix *-ose*. For example, arabinose, raffinose, stachyose, and xylose are all plant sugars. Sucrose, however, is our most widely used sugar in the food supply. Sugar cane, the most abundant source of sucrose, is nature's most efficient solar energy trap. Like all green plants, the sugar cane collects solar energy and by photosynthesis converts it into sucrose. This process occurs deep inside the leaves, in cell parts called chloroplasts. With favorable temperature and moisture, the chloroplasts absorb carbon dioxide from the atmosphere. Sunlight forces the release of oxygen, thereby making it possible to manufacture sucrose, which then is stored in the plant's stem. Each sugar cane stalk contains about 15 percent sucrose.

When we eat sucrose, we reverse the photosynthesis process by converting the sucrose back into energy. Consumed sucrose is digested, converted into dextrose in the small intestine, and stored in the liver as glycogen. Oxygen ignites the energy cycle. Glucose, as needed, is released into the blood stream and metabolized by the body to create energy. The body releases carbon dioxide and water, which can be used by plants to create more sucrose. Thus, the cycle is completed and begins anew.

It is thought that, at least 8000 years ago, sugar cane originally grew in the South Pacific. From there it was transported to India where, for the first time, juice was extracted from the cane and evaporated to a solid state called Khanda (candy) in Sanskrit. Perhaps this product was the first processed food.

Sugar cane was known in China, where the juice was used medicinally by herbalists. By the tenth century, the Persians and Egyptians found a simple method to produce crystallized sugar from the cane. During the Crusades, samples of "honey from reeds" were taken back to Western Europe as a curiosity. Sugar remained an expensive rarity. One of the earliest records of sugar use in Western Europe dates back to the thirteenth century when Henry III requested three pounds of sugar "if so much is to be had" for a notable feast. By the sixteenth century, due to brisk trade, sugar became more affordable and was used as a sweetener in British confection shops. By the seventeenth century, the growing of sugar cane and the manufacture of sugar as well as molasses and rum were well established in tropical countries.

The year 1800 added a new dimension. The British blockade of French ports cut off West Indian cane sugar supplies to France. Napoleon requested French scientists to find a suitable substitute. Many plants were investigated. Benjamin Delassert produced a crystallized sugar from beets and won Napoleon's Cross of Honor. Napoleon acted speedily and energetically to establish several hundred sugar beet refineries to free France from further importation of "English" sugar.

The development of beet sugar challenged cane sugar's pre-eminence and enabled temperate regions to enter the sugar market in competition with tropical countries. As rivalry rose, production increased, sugar prices declined, and the product became accessible to more people.[1]

Both cane and beet sugars have many versatile characteristics and are particularly attractive for food and beverage processing. Sugar acts as a preservative with fruits, jellies, jams, and preserves, and enhances fruit flavors. In products such as canned cherries, sugar makes them plumper, less acidic, and more palatable. Sugar is useful as a curing agent for processed meats. With ice cream, baked goods, and confections, sugar contributes bulk and texture. In soft drinks, sugar gives the liquid "body" that makes it taste heavier and more appealing in the mouth than if the liquid were thinner. In baked goods, sugar serves several functions, by acting as a medium in fermentation, contributing to crust color and flavor, and holding moisture in products, which extends their shelf life.

Perhaps the first observations of human health problems related to high sugar consumption were reported in the first century A.D. by three Hindu physicians, Caraka, Susruta, and Vaghbata. They had observed a disease that at the time was fatal (and is recognized by modern physicians as *diabetes mellitus*) and suggested that large amounts of sugar were bad for such people. Such warnings were not sounded again until the early nineteenth century, by which time the price of refined sugar was within reach of prosperous classes on continental Europe, in England, and in America. For the first time, quantities of sugar were consumed. Between 1850 and 1900, worldwide consumption of sugar increased tenfold; between 1900 and 1940, it tripled again.

In the sixteenth century, a German traveler in England, on seeing Queen Elizabeth, suggested that her blackened teeth might be due to sugar indulgence. Perhaps this was the first recorded observation that high sugar consumption may lead to dental problems.[2] As sugar consumption rose in the developed countries in more recent times, the dental problems became apparent.

While the sugar interests vehemently reject most suspected relationships between high sugar consumption and many human health problems, they grudgingly admit its role in tooth decay.

By far, the most significant human study was done in Sweden, reported in 1954, and known as the Vipeholm Dental Caries Study. More than 400 adult mental patients were placed on controlled diets and observed for five years. The subjects were divided into various groups. Some ate complex and simple carbohydrates at mealtimes only, while others supplemented mealtime food with between-meal snacks, sweetened with sucrose, chocolate, caramel, or toffee. Among the conclusions drawn from the study was that sucrose consumption could increase caries activity. The risk increased if the sucrose was consumed in a sticky form that adhered to the tooth's surfaces. The greatest damage was inflicted by foods with high concentrations of sucrose, in sticky form, eaten between meals, even if contact with the tooth's surfaces was brief. Caries, due to the intake of foods with high sucrose levels, could be decreased when such offending foods were eliminated from the diet. But individual differences existed, and in some cases, caries continued to appear despite avoidance of refined sugar or maximum restriction of natural sugars and total dietary carbohydrates.[3]

Both cane and beet sugars are sucrose. Chemically, their structures are identical and they are used interchangeably. Both are 99.9 percent pure, and chemists affirm that they cannot be distinguished one from the other. Nonetheless, individuals who are highly sensitive to cane may suffer adverse reactions to cane sugar consumption but not necessarily to beet sugar, and vice versa.

Cane sugar sensitivity was reported to the medical community as early as 1925, and beet sugar sensitivity, the following year.[4] The possibility of cane sugar's being an allergen has received scant attention, although it is part of a botanical family of cereals of which various members are common allergens. In addition, cane sugar, consumed frequently in the typical American diet, fulfills another prerequisite as a suspicious food allergen. Documented case studies have been reported in medical journals demonstrating chronic allergic symptoms from cane sugar consumption, as well as from intravenous injection of invert sugar of cane origin.[5] (Invert sugar is a mixture of sucrose and dextrose in special form; see Glossary.)

An extreme example of cane sugar sensitivity was reported in a 55-year-old man, subject to allergic rhinitis, fatigue, headache, and depression. Within a few minutes after an allergy test injection of the usual concentration of an extract of cane sugar, he reported a heavy sensation in his head, yawned several times, and developed facial flushing. These symptoms were followed rapidly by extreme apprehension, crying, and other signs of acute depression, as well as physical symptoms, including chills.[6] While this case may appear extreme, it demonstrates the allergenicity of cane sugar.

Sensitivity to beets as a vegetable has been long recognized, but recognition of sensitivity to beet sugar was not discerned. For a long time, beet sugar was the main source of granulated sugar consumed in certain well-defined geographic areas of the United States, especially where beet sugar mills were located, which were regions relatively distant from cane sugar refineries. The area included the entire western half of the country, and to some extent the north central states, notably Michigan, Minnesota, and Wisconsin. In more recent times, however, regions have not been so well defined. Processors now make their selections based on wholesale prices of beet or cane sugar. Due to this changed practice, beet sugar use in food and beverage processing is no longer confined to certain regions, and products containing beet sugar are in national distribution.[7]

As with cane sensitivity, documented case studies of beet

sugar sensitivity have been reported in medical books.[4] * An extreme example of beet and beet sugar sensitivity was reported in a 32-year-old woman who, in being tested with beet sugar, developed an acute psychotic condition characterized by disorientation, regression, and depression. The condition persisted for three days, followed by residual amnesia and other symptoms. These reactions were recorded in motion pictures and presented before the American College of Allergists in February 1951.[6]

Federal labeling regulations fail to differentiate among the types of sugars used in processed foods and beverages. Obviously, this arrangement is convenient for processors but health-threatening for a segment of the population. The word *sugar* on a food or beverage label gives no inkling as to its origin from cane, beet, or corn and illustrates one inadequacy, among many, of food labeling regulations.[8]

As part of a review of all food additives identified by the term *Generally Recognized as Safe* (GRAS), the Food and Drug Administration organized a task force to evaluate the health aspects of sucrose as a food ingredient. The report, prepared by the GRAS Review Committee of the Federation of American Societies for Experimental Biology (FASEB) consisted of a four-year review of scientific literature. In 1976, the committee concluded that "reasonable evidence exists that sucrose is a contributor to the formation of dental caries," at current use level, and especially with sticky sweets consumed between meals. Also, the committee noted that undesirably high total caloric intake may lead to obesity. However, the committee felt that any link between sucrose and diabetes was circumstantial only, and that

> there is no clear evidence in the available information on sucrose that demonstrates a hazard to the public when used at levels that are now current and in the manner now practiced.

* Note references that appear out of sequence in any chapter refer to previously cited notes within the same chapter.

However, it is not possible to determine without additional data whether an increase in sugar consumption — that would result if there were a significant increase in the total of sucrose, corn sugar, corn syrup, and invert sugar added to foods — would constitute a dietary hazard.[9]

Four years later, at the end of 1980, at completion of the first comprehensive GRAS list review, sucrose was placed in class 2, repeating the same idea that present levels were presumed safe, but further data were needed.[10]

Two researchers at the Carbohydrate Nutrition Laboratory Nutrition Institute, USDA, Drs. Sheldon Reiser and Bela Szepesi, attacked FASEB's conclusions, charging that abundant evidence demonstrates that sucrose is one important dietary factor responsible for diabetes, obesity, and heart disease. Ample data demonstrate the dramatic rise in diabetes incidence that occurs about one generation after a culture begins to consume high sucrose levels.[11] Reiser and Szepesi cited the clear cause-and-effect case of Yemenite Jews who, after relocation in Israel for 30 years, experienced a "dramatic and tragic" diabetes increase resulting from high sucrose consumption. Rise in the diabetes rate is closely followed by a rise in vascular diseases.

Reiser and Szepesi noted that some individuals consume much more sucrose than the average intake. While sucrose can be shown to produce adverse effects, including all the diabetes symptoms in susceptible animals, the same starch level in the same type of animal does not produce diabetes symptoms. In animal studies, virtually all the clinical diabetes symptoms can be produced by feeding high sucrose levels, but not by feeding high fat or cholesterol levels.[12]

The FASEB report acknowledged that a segment of the population appears to have a genetic predisposition to experience large and permanent increase in blood triglycerides (fats) when consuming sucrose-containing diets.[9] In sensitive individuals, these effects have been observed with sucrose levels as low as 20 to 25 percent of the total caloric intake, *levels well within the average sucrose intake range*. This type of elevated blood fats (hyperlipidemia) has been associated with health problems such as abnormal glucose tolerance, diabetes, and heart disease. Therefore, Reiser and Szepesi charged, the FASEB committee should not have concluded that sucrose intake at present levels represents no health hazard since there are approximately 15 million carbohydrate-sensitive adults in the United States. Sucrose, by itself, may be a very important cause of heart disease and diabetes in 10 percent of the population. For the remaining 90 percent, while sucrose, by itself, may not be a *primary* risk factor, by virtue of its synergistic interaction with dietary fats (cholesterol and triglycerides) sucrose must be considered an important risk factor in vascular disease and diabetes development. Therefore, Reiser and Szepesi strongly recommended that sucrose intake from all sources except fresh or processed fruit without added sugar be decreased by a minimum of 60 percent and be replaced by complex carbohydrates from vegetables and cereals. These recommendations reaffirmed those made nearly a year earlier in the *Dietary Goals*.

In order to implement necessary reform, Reiser and Szepesi made several suggestions. A concerted national effort should be launched to identify carbohydrate-sensitive individuals. Sucrose intake in these individuals should be exclusively from fresh and processed fruit without added sugar. For the protection of the remaining population, food and beverage processors should be required to identify the amount of added sucrose on labels. Further, a national campaign should be launched to inform the public about the health hazards associated with high sugar consumption.[12]

Dextrose: corn sugar

More than half of all cornstarch milled in the United States is converted into corn sugar or corn syrups by means of hydrolysis with heat and acid or enzymes. The resulting products include dextrose (glucose), maltose, liquid and spray-dried corn syrups, liquid corn sugar, lump sugar, and crystalline dextrose.

Corn sweeteners have extensive use throughout the food industry, and their low cost, compared to that of sucrose, makes them especially attractive. In addition, they possess special characteristics that favor their use to supplement, rather than replace, sucrose. At two percent concentration, dextrose is only about two-thirds as sweet as sucrose; but as the concentration of sweetness increases, the difference in sweetness decreases. At 40 percent dextrose concentration, it is difficult to distinguish between the sweetness of dextrose and sucrose. Corn syrups also have this characteristic. Hence, dextrose or corn syrups, combined with sucrose, yield a sweeter mix than would be expected. For this reason, processors can replace some sucrose with cheaper corn sweeteners in such products as the syrup used in canned fruit without loss of sweetness.

Another attractive feature of corn syrups is their ability to inhibit crystallization of sucrose and other sugars, of importance in products such as hard candy, jam, jelly, preserves, syrup, and ice cream. Both dextrose and corn syrups lower the freezing point of solutions. Ice cream processors take advantage of this characteristic and use dextrose and corn syrup to replace some sucrose in order to reduce ice crystal formation and prevent sugar crystallization.

All corn sweeteners are readily soluble. Corn syrups add body and texture to soft drinks, and chewiness to confections and chewing gums. Both dextrose and maltose readily ferment yeasts and are used for this purpose by beer brewers, bourbon distillers, and bakers. Also, bakers find that dextrose contributes an attractive brown crust color and flavor to products.

Dextrose and corn syrups can inhibit undesirable oxidation in certain foods and helps retain bright colors in such products as strawberry or peach preserves, cured meats, and catsup. All corn sweeteners highlight color and impart sheen to food surfaces, thus increasing their sales appeal. Corn sweeteners can absorb and retain moisture in products, important for keeping hard candies from sticking or lollipops from dripping.[13]

Glucose is the term for blood sugar in the body. Unfortunately, the same term was applied to the commercially processed sugar that is derived from cornstarch. Confusion has resulted (see Glossary for clarification). The blood sugar, glucose, needs to be maintained at a steady level within the body, and the human organism has mechanisms to sustain an equilibrium. If large amounts of processed glucose, or any other sugars, are consumed, this equilibrium becomes unbalanced.

Food processors began to use glucose in the late nineteenth century as a cheap sweetener and filler and frequently substituted it for sucrose with canned fruit, candies, and other products. Also, commercial glucose was used to adulterate honey and molasses.

Harvey W. Wiley, M.D., working for the Indiana State Board of Health in the late 1800s, became interested in examining food products, especially sugars and syrups, for adulteration. He quickly discovered that "glucose and its near relations have been, are, and will continue to be the champion adulterants."

Wiley considered glucose to be a dangerous sweetener due to its low sweetness level. Large quantities could be consumed in foods with little recognition of its presence. Glucose could replace nutritious dietary components to an even greater degree than sucrose. For corn-sensitive individuals, substitution of glucose for sucrose, with incorrect label information, was not only deceptive and an economic cheat, but hazardous. Yet even today, FDA continues to seize mislabeled products containing glucose as an adulterant.[14]

In 1902, glucose was introduced into the retail market, but

the public erroneously believed that the product was made from glue. To overcome this prejudice, the product was renamed "corn syrup." Wiley, who by this time headed the federal agency that was the forerunner of FDA, charged that the term *corn syrup* made the product misbranded and subject to seizure. A food was considered misbranded if it bore the name of another substance. It happened that an earlier product existed that was a true corn syrup, made from pressed concentrated cornstalk juice.[15]

Hearings were called before the Board of Food and Drug Inspectors, and the board decided that *corn syrup* was an inappropriate term for the product. But the corn refiners brought heavy political pressures on the federal regulatory agency, on congressmen, and on President Theodore Roosevelt. The refiners launched an organized campaign that, in retrospect, appears like a harbinger of actions taken several decades later with cyclamates and saccharin. Individuals and trade groups were organized in massive campaigns for resolutions and protest letters to flood Washington, D.C.

A new hearing was called. The refiners filed an impressive compilation of opinions from prominent chemists. Wiley charged that college and university scientists were paid to sign testimonials favoring the term *corn syrup*, without understanding the issue of food adulteration and misrepresentation. The campaign succeeded. The government reversed its earlier decision and accepted the term. By February 1908 the issue was closed and Wiley lost his long hard struggle to keep glucose out of the American food supply.[16]

Sugars of corn origin are allergenic for many corn-sensitive individuals. As early as 1935, corn's allergenicity was reported to the medical community. At that time, corn ranked fourth among various foods listed in the order of incidence of sensitivity.[17] A decade later, in a study of 160 adults and 40 children, corn ranked first among common food allergens, due to the increased and repeated consumption of corn and corn constituents in the food supply.[18] In 1944, the medical community was alerted to recognize allergic reactions in corn-sensitive individuals induced by a variety of foods including bacon, candy, gum, ices, and commercially canned fruit, as well as corn derivatives in pharmaceutical preparations.[17]

Among six documented cases described in 1949, corn sugar used in sugar-cured ham was sufficient to induce symptoms of chronic fatigue, depression, melancholia, irritability, rapid heart beat, intermittent chills, generalized muscle aches, chronic sore throat, postnasal discharge, intense itching of the eyelids and ear canals in one corn-sensitive patient. The other five individuals revealed different symptoms that were also wide ranging.[18]

Glucose syrups in feeding formulas were reported to cause diarrhea in most normal babies, not only corn-sensitive ones. Corn sugar in infant feeding formulas can produce corn-allergy symptoms of eczema and gastrointestinal upsets in corn-sensitive babies. Corn syrups and malted corn products used as carbohydrate sources for solid foods intended for babies also induce allergic reactions, and repeated corn-containing foods and beverage exposures may continue with older children, and into adult years. [19]

Theron G. Randolph, M.D., a pioneer in food allergy, noted that for many corn-sensitive individuals, chronic symptoms fail to subside until all corn sugars are eliminated totally from the diet. Randolph suspects that the practice of feeding dextrose intravenously frequently induces reactions in hospitalized corn-sensitive patients.[20]

Not only is corn sugar particularly effective in carrying all the allergenicity of corn, but certain corn-sensitive patients tend to react in tests with corn sugar consumption more quickly than to other fractions of corn. Reactions to corn sugars such as dextrose (glucose) and dextrin were observed within 10 to 12 minutes; to cornstarch, within 20 minutes; and to corn oil, from 2 to 3 hours. [6]

Corn-sensitive individuals need to avoid saccharin sweeteners that are mixed with dextrose; certain granulated sugars labeled "superfine, pure cane granulated sugar with dextrose"; talc-coated rice, to which glucose is added to make the rice look shiny; sorbitol and mannitol, sweeteners generally made from dextrose; and confectioner's sugar that may have corn-starch added to prevent caking.

In 1960, our annual consumption of corn sweeteners was 11.6 pounds dry basis per person, which consisted of 3.4 pounds of dextrose and 8.2 pounds of traditional corn syrups. At that time, high-fructose corn syrups (HFCS) did not exist. By 1980, our yearly consumption of corn sweeteners had nearly quadrupled, at 41.1 pounds per person. The late entry, HFCS, had a dramatic impact. While dextrose remained relatively stationary at 3.8 pounds, and traditional corn syrups had more than doubled at 18.4 pounds, HFCS, the newcomer, accounted for 18.9 pounds. [21]

The greatly increased use of corn-derived sweeteners has made corn sugar allergies more common than ever. Physicians working with food allergic patients find that the incidence of diagnosed allergy to major foods is approximately as follows: Among the first three are corn and corn sugar; among the first six, beet and beet sugar; and among the first twelve, cane sugar. While some cases are identified, it is suspected that many more remain undetected. [6]

Some of the adverse health effects from high levels of sugar consumption were noted in 1977 by Dr. Herman Kraybill of the National Cancer Institute. In humans, high sugar intake replaces nutrient-rich foods with empty calories. It contributes to the formation of dental caries. It may affect the processes of growth and maturation. If the metabolism of an individual is faulty, high sugar intake contributes to overweight, can cause and aggravate diabetes, and can cause and aggravate low blood sugar and the allergies associated with this health problem. High sugar intake increases the risks of diabetes and pancreatic cancer. It is implicated in high fat levels in the blood (both cholesterol and triglycerides). It is a suspected contributing factor in the development of atherosclerosis.

In addition, Kraybill cited possible adverse health effects in humans, based on animal experiments. High levels of sugar consumption increased the demands on enzyme systems of the liver, kidney, and serum in test animals. High levels of sugar consumption impaired glucose tolerance, speeded up kidney disease, and shortened the life span.[22]

Traditional Sweeteners: Overrated?

Raw and brown sugars, molasses, honey, maple syrup, sorghum, malt, grain syrups, whey

No one form of sugar — whether honey, granulated white sugar, brown sugar, or molasses — is any better than another, according to Kay Munsen, Iowa State University extension nutritionist. When food experts talk about eating more "naturally occurring" sugars, she explained, they are not talking about less processed forms of sugar like honey. They mean sugar already present in fruits and vegetables.
— *Chicago Tribune,* February 15, 1979

WHEN THE WORD *sweeteners* is used, most people think of raw, brown, and white sugar, or syrups like molasses, honey, maple syrup, or sorghum. These common sweeteners are familiar; some have had a long tradition of use. Erroneously, many people regard these sugars as "natural." They are natural only in the sense that they are derived from plants. But all have been extracted and concentrated, and it is correct to regard them as processed sugars. The term *natural sugars* should be limited to those sugars present in fruits and vegetables, which have not been extracted and concentrated. How desirable are the traditional but processed sweeteners?

Raw sugar

At a "sugar centrale" the juice is squeezed from sugar cane by means of heavy rollers, then filtered and boiled in a huge vacuum pan to make the crude sugar crystallize. The newly formed crystals are centrifuged to separate them from the molasses, which is the remaining liquid that did not crystallize.[1]

At this stage, the crystals are "raw sugar." Being coated with molasses they are brownish and sticky. These crystals are shipped to a refinery for further processing. The film clinging to the crystals is removed and the resulting product is refined white cane sugar, also known commonly as table sugar.

Meanwhile, back at the sugar centrale, the molasses is processed further, and this time the crystals formed range from yellow to brown, with a distinctive taste. These are the so-called "yellow sugar" and "brown sugar." Viewed microscopically, these sugars are still white inside, with molasses coating. These sugars clump due to the sticky molasses. As consumers know, the molasses coating dries out after prolonged storage of these sugars, and they harden.

The remaining molasses is processed still further, as many as seven or eight times, to extract as much crystallized sugar as is feasible. With each processing, the molasses becomes progressively darker. After as much crystallization as possible is obtained, the process ends. The remaining syrup is blackstrap molasses.[2]

Reputable writers who discuss natural foods generally warn about raw sugar's shortcomings. Many natural and health food stores refuse to stock raw sugar or products made with it. Nevertheless, a persistent myth continues that somehow raw sugar is superior to refined white sugar. In part, this myth has been perpetuated by processors and sellers of raw sugar. Recently, restaurateurs have been urged, by advertisements in their trade journals, to put raw sugar on tables. The product "says you're aware of natural food trends among today's health conscious consumers" and offers "new texture, new and more natural color." The packets are labeled "unrefined, natural sugar."[3] Fact or fiction?

So-called raw sugar offered at retail markets as well as to restaurants is not the raw sugar consisting of newly formed crystals centrifuged and separated from molasses. While this type of raw sugar is sold in some countries, FDA considers it unfit "for direct use as a food or ingredient because of the impurities it contains." FDA forbids its sale, and rightly so. At this stage, raw sugar may be contaminated by particles of sand, earth, molds, bacteria, yeasts, sugar lice, fibers, lints, and waxes.[4] Some samples of such sugars have been found to contain up to three percent of such undesirable extraneous substances.

If raw sugar is washed once by centrifuge under careful sanitary conditions, some but not all of the dirt will be removed, along with some solid matter and bacteria. If the resulting product meets the minimum sanitary level established by FDA under its Filth Tolerance Allowances the sugar is cleared for sale. This sanitized product is known as turbinado sugar.[5] According to The Sugar Association, Inc., contaminants may be found in samples of turbinado sugar.

Raw sugar, as produced in this country, actually is white sugar with traces of cane or beet pulp added back to give it approximately the same appearance and taste as genuine raw sugar.[6] Or, depending on the degree of coloring desired, molasses may be added back to refined white sugar, ranging from 5 to 13 percent of molasses.[3] A special crystallization process is used for some light-colored sugars, designed to create the illusion of rawness. Whatever iron content is present is mainly from contamination by tiny particles worn off the machinery in which the sugar is prepared from the cane. [6]

In 1972, the Federal Trade Commission (FTC) took action against a company producing a raw sugar product, charging that the company engaged in false advertising. FTC based its actions on the following: The product was not an organically grown but a processed food; it did not have unique qualities unavailable in other types of sugars; it was not a significant source of vitamins and minerals, nor a significantly greater source than refined sugar (neither contains nutritionally significant amounts); the absence of chemicals and preservatives did not make the product substantially different from, or superior to, other sugars. [7]

Nor does any justification exist for the current claim "unrefined, natural sugar" on raw sugar labels. Such products have undergone several refining processes, including use of various substances such as lime, phosphoric acid, charred bones, diatomaceous earth, and carbon, for purposes of clarification, adsorption, and whitening.

Brown sugar

In countries where cane is grown, some brown sugars are made simply by cutting short the refining process so that the sugar retains some of the molasses that otherwise is washed out or removed in the final refining treatment. At times, brown sugar is made simply by adding brown caramel coloring to white sugar, a highly deceptive practice. [4]

The technique for brown sugar production in this country makes the product *more* refined than white sugar, and the process may offer health hazards. The raw sugar is washed to remove the molasses coating, and centrifuged to separate the crystals. Then it is heated to a melting point, filtered, and "decolorized" with animal-bone charcoal. With repeated boilings, the decolorized liquid becomes crystallized and concentrated. The "brown" color in this sugar results from the bone-charcoal treatment rather than from molasses residue. [8]

In reference to this decolorizing process, the cancerologist,
Wilhelm C. Hueper, M.D., reported,

> In themselves, sugars may not be carcinogenic — but carcino-
> genic impurities may be introduced into sugars when concen-
> trated sugar solutions are filtered for decolorizing purposes
> through improperly prepared charcoal containing polycyclic
> hydrocarbons. Chemicals of the dibenzanthracene type are
> eluted [washed out] from charcoal by concentrated sugar solu-
> tions ... Traces may be introduced in this manner and remain
> in apparently chemically pure sugar.[9]

In a baking trade journal, bakers were advised that

> substantial dollar savings are now available to bakers using
> brown sugar by "making their own." The concept involves
> simply mixing granulated [white] sugar with a special brown
> sugar molasses product. Brown sugar use is increasing sharply
> because of the consumer trend toward natural foods.[10]

Commercial bakers also are offered imitation brown sugar
in liquid form. The product is intended to fortify regular sugar
or to replace up to 100 percent of brown sugar flavoring.

In 1975, a granulated brown sugar product was introduced
in retail markets, described as a "low-calorie sugar substitute"
that "looks, tastes, and smells like old-fashioned brown sugar."
The product consisted of brown sugar and saccharin. [11] How
accurate was the description? "Sugar substitute" described the
saccharin portion correctly, but "low-calorie" was inaccurate
for brown sugar. A cup of packed brown sugar has 821
calories; table sugar, 770. In FDA's quest for greater accuracy
in food labeling the agency announced in 1978 that products
may be labeled low-calorie only if they contain no more than
40 calories per serving. [12]

No brown sugars contain any worthwhile amounts of nutrients. Nor do they contribute any significant amounts of
vitamins or minerals to meet the body's needs.

Molasses

Molasses played a significant role in American history and caused John Adams to call it "an essential ingredient in American independence." Columbus had introduced sugar cane and its resulting molasses into Santa Domingo in 1493. By the eighteenth century, molasses, being the instrument that balanced accounts with the mother country, was regarded as the life blood of colonial commerce. Exports from the colonies of lumber, fish, cotton, and other agricultural staples failed to equal the value of manufactured goods imported from England, until the molasses trade tipped the scale.

Sailing vessels shipped colonist products to the West Indies and returned with molasses, which the colonists fermented and distilled into rum. Part of the rum was consumed locally, but much of it went to Africa in exchange for slaves. Ships returned to America via the West Indies, where some slaves were debarked and promptly bartered for more molasses. The remaining slaves were shipped, along with molasses, to the southern colonies.

England attempted to discourage this arrangement by passage of the Molasses Act of 1733, which levied heavy import duties. By 1764, England proposed more stringent provisions, and molasses, as well as tea, became a precipitating factor in the American Revolution.

In areas along the Atlantic seaboard where molasses imports were substantial in colonial days the custom of molasses use still prevails. Maine and North Carolina lead all other states in current molasses consumption.

The quality of the molasses depends on the sugar cane's maturity, the amount of sugar extracted, and the extraction method. Quality also depends on whether the processor intends to produce molasses or sugar as the primary product. The three major types of molasses are unsulfured, sulfured, and blackstrap products. Unsulfured is made from the juice of 12-

to 15-month sun-ripened West Indian cane. The harvested cane is crushed and heated in clarifying kettles. The juice then flows through a series of copper kettles, becoming concentrated to the desired density. Then it is transferred to tanks where the molasses ages. This process has been used for more than two centuries, but today's machinery insures uniformity and sanitation in the product. The molasses is shipped by barrel or puncheon to warehouses in this country and then is blended. Since the character and flavor differs from each island, blending produces a uniform product. Some molasses may be kept in the warehouse for a year or two for further aging and flavor improvement.

Sulfured molasses is a by-product of sugar refining. The efficient modern sugar plants, by extracting a high sugar level from the original cane juice, yield molasses of lower quality. Most sulfured molasses is made from cane grown not in the West Indies but in areas where the ripening season is not long enough to allow cane to mature. The green sugar cane is treated with sulfur fumes during the sugar extraction process, and as a result, sulfur residue remains in the molasses.

"First centrifugal" and "second centrifugal" are two types of sulfured molasses. The former is sugar cane juice boiled to the proper density in vacuum pans and then centrifuged to extract sugar crystals. A yellow-colored molasses remains and is subjected to a second boiling and centrifuging, extracting additional sugar crystals. The remaining molasses is darker and contains a larger percentage of natural gums and ash, and has a less agreeable flavor.

Blackstrap molasses is the waste product of sugar processing, usually sold to produce industrial alcohol for rum and yeast, and for use in animal feed. This molasses results from a third boiling, with more sugar crystals having been extracted. The remaining product, from which it is no longer profitable to remove more sugar, is blackish in color, with a burnt, bitter taste. If the sugar refinery is imagined as a laundry where crude sugar is washed as clothes, blackstrap molasses would be equivalent to the wash water that carries most of the mineral matter and all of the gum, ash, dirt, and indigestible matter present in the unprocessed crude sugar.[18]

Food processors use molasses for its special characteristics in addition to its sweetening ability. Molasses has the ability to mask certain unpleasant flavors, such as the bitter taste of bran in bread. Molasses can mimic related flavors, such as chocolate, butterscotch, coffee, and maple, and may be added to foods and beverages with those flavors. Molasses is effective in controlling the water activity of foods and appears to have a natural anti-oxidant component that helps retard food spoilage. [14]

Many food processors use liquid and dry molasses blends. Such products are not 100 percent molasses. One light-brown blend consists of molasses, wheat starch, and soy flour. The total sugar content is 60 percent, consisting of 30 percent sucrose and 30 percent other sugars. One medium-brown blend has a total sugar content of 53 percent: 37 percent sucrose and 16 percent other sugars.

One free-flowing granulated molasses blend consists of 75 percent molasses and 25 percent corn syrup solids, while another has 75 percent molasses and 25 percent soy bran and cornstarch. Such blends are used for many processed foods, including breads, cookies, pies, health foods, mixes, cereals, and sausages. The strong molasses flavor acts as a masking agent with some of these products.[15] While the ingredients in these blends may be acceptable food components, the word "molasses" on food labels is misleading. Consumers are unaware of the presence of corn, wheat, or soy constituents, any of which may be hidden allergens.

Is molasses a superior sweetener? Molasses, primarily a carbohydrate, provides energy by supplying calories from its sugars. Molasses is about 73 percent sugars, mainly sucrose, dextrose, and levulose. Molasses contains some nutrients, notably iron, potassium, calcium, and phosphorus, and some trace elements, especially zinc, copper, manganese, and chromium. But are these nutrients presents at levels that make significant contributions to the daily diet?

One molasses producer stated that "the natural sugar in molasses is taken up by the body without any undue strain and can be used in practically any quantity." This statement may sell molasses but is poor advice. The producer continued, "only liver compares favorably with molasses as a source of iron, and molasses rates even in this comparison [and] as a less expensive source." [13] Untrue. One three-ounce serving of calves' liver provides 12.1 milligrams of iron; less expensive pigs' liver, 24.7. One would need to consume a 12-ounce bottle of light molasses, which contains 21.2 milligrams of iron, to approximate the amount in pigs' liver; a 12-ounce bottle of medium-colored molasses supplies 29.6. [16] The liver provides many other nutrients, in addition to iron, at high levels. But to use molasses as a dependable iron source, one is forced to consume an undesirably high sugar level.

Is blackstrap molasses better? A mystique has evolved and been perpetuated about its benefits and nutrients. Sulfur and blackstrap molasses, used as spring tonic and as a pimple lotion, are well-established rites. Folklore provides the notion that blackstrap molasses, mixed with cinders, enriches the blood; with goose grease, cures croup; and in the North Carolina hills, a mixture of blackstrap molasses with whiskey and linseed oil purportedly relieves whooping cough. Blackstrap molasses enthusiasts have termed it a "miracle" food.

Due to the concentration of blackstrap, it contains similar nutrients found in other molasses, but at higher levels. Compared to a three-ounce serving of pigs' liver, a 12-ounce bottle of blackstrap molasses provides 79.4 milligrams of iron. While this is a significant level, the bitter unpalatable nature of blackstrap molasses makes it self-limiting. And, it is still lots of sugar. Also, according to Dr. Roy E. Morse of the Food Science Department, Rutgers University, there's a lot of iron in [blackstrap molasses], just as the health food people claim, but there may also be arsenic and some other unpleasant things, which have become more concentrated with each run. The possible dangers from these components have caused the FDA to forbid the sale of true blackstrap as human food.[1]

Blackstrap molasses sold for human consumption contains about 30 percent sucrose. For food preparation, blackstrap molasses is more palatable if blended with sweeter substances, such as other molasses or honey. Due to the stickiness of all types of molasses, they cling to the teeth. If not removed promptly, residues can lead to dental decay.

Prudence dictates that molasses should be used sparingly, as a flavoring agent rather than as a nutrient. Because molasses is mainly sugars and highly concentrated, the use of molasses needs to be limited.

Honey

Numerous books and articles have been written in praise of honey. Among the earliest of sweeteners, through many centuries, legends and folktales about honey have evolved in mythology, religious ritual, medical practice, and social custom. The high esteem for honey is reflected in expressions such as "nectar of the gods," "youth elixir," or "molten gold." Numerous expressions in language incorporate the word honey to express pleasant tastes, and other attributes.

Today, honey's popularity is understandable. The term *natural* is positive; *synthetic,* negative. Similarly, *unprocessed* is preferable to *refined.* Synthetic sweetener safety is doubtful, and refined sugars are implicated in dental decay and other health problems. Honey symbolizes natural and unprocessed, with implied attributes of healthfulness and safety.

Nectar gathered by bees consists of more than 80 percent sucrose, which bees are able to metabolize by means of their digestive enzyme, invertase. In the time elapsing between gathering the nectar and returning to the hive, the bees have split the sucrose of the nectar into two simple sugars, glucose (dextrose) and fructose (levulose). This mixture, invert sugar, [17] is sweeter than the original nectar from which it is derived. Invert sugar, when eaten by human beings, provides a direct energy source since it requires no digestion. Fructose, highly soluble, does not crystallize readily, but glucose does. The proportions of fructose to glucose vary in different honey varieties. For example, tupelo honey, from the gum tree flower's nectar, has a high fructose content and resists crystallization.[1]

Honey is prized, however, not so much as a source of ready energy but for its flavor. Dissolved aromatics give each honey its own distinctive flavor and appearance. The remaining components in the mixture are mostly water, a small amount of sucrose, dextrin, gums, and a modicum of minerals, vitamins, enzymes, and bee pollen. The presence of these ingredients, albeit in small amounts, is the prime reason for honey's being so highly valued by persons interested in natural/health foods.

Compared to other caloric sweeteners, honey consumption in the United States is minor and remains stable. In 1963, the average American ate more than 107 pounds of cane, beet, and corn sugars annually, compared to only 1.1 pounds of honey. By 1980, while cane, beet, and corn sweetener consumption had risen to 127.3, honey declined slightly to 1.0.

As honey progresses, from beehive to table, can it be considered as "unrefined"? First, honey needs to be extracted from the hive. Cottage-industry beekeepers may use a simple bee-escape device to remove bees and gently brush off any stray ones with a soft brush. To save time, large-scale packers may resort to toxic materials such as carbolic acid, nitrous oxide, proprionic anhydride, and benzaldehyde. Reportedly, these substances dissipate quickly and leave no residues. However, repeated exposures to these substances may shorten the bees' lifespan. [18]

Honey labels that state "unheated," "no heat used," or "unpasteurized" are misleading. Generally, home packers apply heat to thin honey for easier extraction, to retard crystallization, and to destroy honey yeasts that cause fermentation and spoilage. The conscientious apiarist will extract with mild heat (only up to 120°F.) and such honey will retain its small amounts of nutrients. Studies showed that filtered honey heated to 150°F. lost 27 to 30 percent of its thiamin; 22 to 45 percent, riboflavin; 8 to 22 percent, pantothenic acid; 15 to 27 percent, niacin; and 9 to 20 percent, ascorbic acid. [19] Prolonged exposure to heat causes enzyme and protein breakdowns as well as color and flavor impairment. Large-scale honey packers may flush-heat honey briefly through heated pipes to above boiling temperatures.

"Buttered" or "creamed" honey is finely granulated, but this appearance is no guarantee that the honey has not been heated. The honey may have been heated, then recrystallized by seeding it with honey crystals to produce granulation.

Extracted honey is strained or filtered to remove impurities and to clarify the product. Some people look for label terms "unfiltered" or "unstrained" honey, for a small amount of pollen still present. Along with the pollen, present in insignificant amounts, may be other particulates such as bees' wings and legs.

Filtration is used to grade honey; the greater the clarity, the higher the grade. "Fancy" U.S. Grade A, the top rating, is given to honey strained through a screen or cloth of at least 90 meshes per square inch, a gauge that approximates that of nylon stockings. "Choice" Grade B is strained through a coarser sieve, 57 to 79 mesh; "Standard" Grade C, 18 mesh; and "Substandard" Grade D, less than 18 mesh. This USDA grading, used by most states, has no significance other than as a measurement of particle size of the extraneous matter filtered out of honey, and is *unrelated to quality*.

Actually, USDA uses a rarely publicized quality rating, with three factors on a 100-point scale. A maximum of 50, 40, and 10 points is given respectively for flavor, absence of particles, and clarity. Obviously, flavor is more important than clarity. Neither official grades nor quality ratings carry any marketing constraints, and legally all four grades are salable.

The cottage-industry apiarist simply may strain the honey through a filter, or just allow it to settle overnight. Large-scale honey packers may mix honey with diatomaceous earth to remove particulates as the honey is pumped through a series of filtration discs.

As honey ripens in the hive, bees reduce its moisture content to less than 17 percent, which results in honey with a smooth, mellow flavor. The glucose in the honey releases hydrogen peroxide, which retards fermentation. At times, however, beekeepers who are eager to market honey early, extract unripe honey prematurely. Such "green honey," with moisture frequently above 20 percent, has a disagreeable biting flavor. Its low glucose concentration reduces its resistance to bacterial contamination, and this honey readily ferments and spoils.

Legally, honey is defined as "the nectar of the floral exudations of plants gathered and stored in the comb of honeybees." [20] One writer gave false assurance that "the surest way to get untreated, unheated honey is to buy it in the comb." [21] Commercial comb honey is not necessarily unheated and unfiltered. Such products may consist of processed honey that, along with clean empty honeycombs, is placed in containers.

A food trade journal described honey labeling as a "neglected problem" by the natural food industry, whose honey sales constitute a significant segment of the honey market. Terms such as *natural, old-fashioned, country-style, organic,* or *undiluted* on honey labels are meaningless. One honey processor justified the term *organic* on the label by citing one dictionary definition of organic as meaning "free from foreign particles." *Undiluted* would mean that no water has been added to the honey; if it were, the product would be considered adulterated and subject to seizure.

Label information about the honey's source may be imprecise. Honey labeled as the product of a single source, such as sage, clover, orange blossom, or buckwheat, may in fact be blends of more than one source. California and other states require that only 51 percent of the product be derived from one source in order to use that source name on the label.

Honey blending is practiced to alter moisture content, improve flavor, or retard granulation. For example, the bitterness of orange blossom honey may be mellowed by blending it with a milder, lighter honey.[22] Buckwheat honey is so strongly flavored that in many states it may comprise only 10 to 15 percent of a product labeled "buckwheat honey." [23]

The Florida Department of Agriculture has established more stringent controls. Identifiable types or varieties of honey must be made entirely from the stated origin. Florida issues official tags or labels to mark containers "certified tupelo honey," "certified orange blossom honey," or others. Regarding honey blends, the Florida Department of Agriculture cautioned Florida apiarists that

> blending honey has reference only to the very best honey and not to any of inferior quality. A poor grade should never be in a blend, or it will ruin all. It is better to put the cheap honey up separately and sell as such. This applies to both the color and flavor of honey. Some poor honey has a fine color, and some very fine honey has a poor color. It is seldom, if ever, advisable to blend dark honey with light, or honey of poor flavor with that of good flavor, but a blend should always be with honey of similar color and quality of flavor.[20]

Blended honey is not necessarily inferior to single-source honey. But the label is inadequately informative to meet the needs of individuals who are sensitive to certain substances that may be present as hidden allergens.

Unfortunately, some honey is adulterated, and this practice is not new. As early as the 1830s, Frederick Accum, a pioneering muckraker of food adulteration, wrote that "honey is frequently adulterated with starch, sand, flour, etc.; for the purpose of improving its colour, or increasing its weight." Accum described his technique for sand detection by dissolving honey in water and allowing any sand that might be present to sink to the bottom.[24]

At present, honey adulteration continues, not with sand but with cheap sugars or syrups. Frequently FDA announces seizures. Since the development of the industrial sweetener, high-fructose corn syrups (HFCS — discussed later), honey adulteration from this source became common and difficult to detect. Until 1978 the available testing techniques were inadequate, but the adulteration grew to such proportions that the practice was an economic threat to both consumers and honey producers. Honey is in limited supply and is costly; HFCS is plentiful and inexpensive. In 1978, the Eastern Regional Research Center, USDA, perfected testing techniques capable of detecting HFCS in honey as well as in bee feed. In either case, such honey is regarded as adulterated and subject to seizure.[25]

In contrast, another USDA branch's efforts are expended in the opposite direction. Researchers at the Bioenvironmental Bee Laboratory, USDA, have simulated the aroma and flavor of bee pollen with a mixture of whey, yeast, soybean flour, dry skim milk, cottonseed meal, and corn gluten. The mixture, encapsulated in starch, is fed to honeybees. The colonies fed this fabricated substitute food yielded as much honey and reproduced as readily as pollen-fed colonies.[26] This type of research should be discouraged. Among other undesirable outcomes would be lack of crop pollination, and potential allergic food reactions to components in the bee feed induced in individuals who normally tolerate honey derived from flower nectar.[27]

Food processors use honey in liquid and dry forms, and such products are not 100 percent honey but rather honey blends containing other ingredients. One product contains only 60 percent honey, with added wheat starch and soy flour as drying aids and calcium stearate to keep this dry form of honey from caking. Another is a blend of 60 percent honey and invert sugar. One advertisement for a blend, offered to food processors, promised savings of up to 30 percent. Such products are used in many baked goods, breakfast foods, snacks, soups, sauces, sugar substitutes, and in a variety of canned, frozen, and tableted foods. Food processors are advised that "use of the product in most cases requires no change in labeling statement," an indication that consumers are misled. When the term *honey* appears as an ingredient on food labels, buyers do not expect that invert sugar is mixed with honey, or that 40 percent of the honey consists of other ingredients. [28] In some instances, health food outlets have sold, in good faith, products that were labeled in this manner and later were found to contain sucrose in the carrier base.[29]

In addition, imitation honey flavors are available to food processors that "look, taste, and perform like honey." Some are created by adding beta-phenylalanine to a sugar such as glucose and allowing it to react at high temperatures. Such products are designed as honey replacers in meat and dairy products, jellies, jams, cereals, baked goods, toppings, glazes, ice cream, candies, and other products in which honey is used as a sweetener or flavoring. These replacers are low in cost, will not crystallize, and store indefinitely. Some are a hundredfold greater than honey in their sweetness intensity and can be diluted with invert sugar syrup. [30]

How safe is honey? Generally, honey is a safe food. However, there are a few practices worth noting. Sulfa drugs and other antibiotics may be used to control bee diseases in the hive. These medications are fed in sugar syrup before or after a honey flow. Empty combs may be fumigated with para-dichlorobenzene (moth balls) or methyl bromide. Either substance may contaminate and impair honey. Many apiarists, concerned about their bees and their own safety as well as the honey quality, refrain from using these adjuncts.

Pesticides may contaminate honey but the problem is of a lesser order than with most foods. Worker bees probe deeply into blossoms where pesticide concentrations are lower than on the outside parts. Pesticide-exposed bees are apt to perish before they reach the hive, and those who do return exhibit such radical behavioral changes that they are refused admittance by the colony.[31]

However, the honey from certain plants can be toxic. Nectar from rhododendron, azalea, oleander, and mountain laurel contain highly cardioactive glucosides. Palmetto and black locust honeys also may be toxic. Some honey contains substances capable of inducing liver cancer and birth defects. [32] Tansy ragwort, a common western weed, contains alkaloids known to inflict liver damage in animals. Bees forage on ragwort during dry summer months. Researchers from Oregon State University found significant quantities (up to 3.9 parts per million) of the ragweed alkaloids in four Oregon honey samples. Heavily contaminated honey, being bitter, is unlikely to be sold uncut, but frequently bitter honey is blended with milder ones. Even trace contamination of toxic ragweed alkaloids is cause for concern.[33]

Honey poisoning has long been known, having been described in 400 B.C. by Xenophon. Near Trebizond, Turkey, his troops found beehives and

> all those who ate the honey went out of their senses and vomited and purged and not a man of them could stand on his feet [recorded Xenophon]. If they ate only a little they seemed like drunken men, if they ate much, like madmen; some even died of it. So they lay in heaps as if there had been

a rout, and they were very unhappy about it. Next day no one died, but about the same time of day they came back to their senses; in another day or two they got up dazed as if they had been drugged.[31]

In modern times, honey poisoning in Turkey is still common, and recognized. Elsewhere, less common, the slight effects experienced from eating a small quantity of toxic honey probably are not well identified. Several outbreaks were reported in New Zealand, attributed to a toxin in the nectar from the tutu plant. Toxins have been detected in U.S.-produced honey, exported to Great Britain.[34]

Poisonous honey, which should be recognized as a problem, is fortunately not widespread. Bees gather nectar from a limited radius of approximately two miles. Generally, bee-keepers are familiar with the flowers in their vicinity, and marketed honey is apt to be produced from safe flower sources.

Raw honey poses a distinctly different problem. Many people choose raw honey for its fine flavor, presence of enzymes, and lack of processing. For most of the population, raw honey is perfectly safe. However, raw honey may be hazardous to very young infants.

Raw honey may contain *Clostridium botulinum,* the bacterium whose spores are responsible for botulism. In 1976, infant botulism was viewed by the Center for Disease Control, U.S. Public Health Service (USPHS), as "a sizable problem" that has gone unrecognized. Infant botulism differs from the botulism that poisons adults, which follows the consumption of toxins created in improperly canned or cooked food. Infant botulism is from *C. botulinum* spores that are widespread in soil, dust, and raw agricultural commodities. These can be swallowed by infants, multiply in the gut, and produce toxins that cause botulism. Infant botulism became one more factor among possible causes of sudden crib death syndrome. However, unlike adult food botulism, which results in about

60 percent mortality in all cases, the infant disease, while critical, generally is not fatal.[35]

In July 1978, the Sioux Honey Association of Sioux City, Iowa, the world's largest honey-packing cooperative, issued a press release, stating that honey and other raw agricultural products fed to infants less than 26 weeks old could produce botulism poisoning. This statement was prompted by epidemiological and laboratory findings at the California Department of Health Services. A two-year study showed, by 1978, that of all foods tested that had been fed to babies who developed infant botulism, honey alone was found to contain *C. botulinum* spores. Of more than 60 honey specimens tested in California, about 13 percent were contaminated. These findings were confirmed independently by four laboratories in other sections of the country, and were not unique for California honey. Although it was believed that honey might account for less than a third of all California infant botulism, the condition involved many other risk factors. Also, any raw agricultural produce, including all fruits and vegetables, can be sources of *C. botulinum*. The California researchers concluded that

> since honey is not an essential infant food we concur with the recommendations of the Sioux Honey Association that honey not be fed to infants under one year of age. Our research on infant botulism casts no doubt on the safety of honey as a food for older children and adults.[36]

Why is raw honey safe for older children and adults, but not for young infants? For older children and adults, botulism is a food poison, which is often fatal. In young infants, however, botulism is an infectious disease, but rarely fatal. The botulinum spores, ingested by the young infant, can grow in the intestine and produce enough toxin to make the infant quite ill. Although all young infants are exposed to botulinum spores from various sources, not all infants become infected.

Susceptibility is believed to be dependent on the composition of the bacterial population in the infant's intestine.

Is honey a superior sweetener? Many qualities are attributed to honey, among which are its easy digestibility, antiseptic and laxative qualities, and medicinal uses. While these virtues may exist, they need to be separated from any discussion of honey's nutritional value.

Tomb reliefs dating from the third millennium B.C. depict Egyptians collecting wild honey by smoking bees from their nest. Probably honey was used far earlier. Since wild honey was found infrequently it was eaten infrequently, and therein lies the lesson. Honey, used infrequently and in small amounts, is an acceptable sweetener. If it is regarded as a flavoring, rather than as a sweetening agent, so much the better. Unfortunately, many people think that since honey is a natural sweetener it can be used in any amount without creating problems and will, in addition, provide ample nourishment. One writer extolled,

> It contains many minerals, such as copper, iron, manganese, silica, chlorine, calcium, sodium, potassium, phosphorus, magnesium. Dark honey is said to have a higher mineral content than light honey. Honey that has not been filtered contains vitamin C from its pollen.

The writer implies that honey is highly nutritious but fails to caution that these nutrients are supplied at exceedingly low levels and are undependable sources. For example, one would need to consume about 230 tablespoons of honey daily to meet the minimum requirement for riboflavin. Similarly, honey's vitamin C content is minuscule, and far below one's daily requirements.

Honey used sparingly as a flavoring agent may be marginally superior to other sweeteners. Honey contains more monosaccharides, which are sweeter than disaccharides. Honey is about 1.3 times sweeter than sucrose, so less can be used.[37] Honey is utilized quickly. Both glucose and fructose have been predigested in honey, so it goes directly into the blood stream for "quick energy." This feature of easy digestibility is unfavorable, however, for hypoglycemics and diabetics. However, if a person is allergic to corn, cane, or beet sugars, honey may be a non-allergenic alternative.

Due to honey's stickiness, it contributes to dental decay. The mouth needs prompt cleansing after honey consumption.

The conclusions are that honey, used sparingly, is an acceptable flavoring or sweetening agent; but as a nutrient source, it is overrated.[38]

Maple syrup

Maple syrup, like honey, is perceived as a natural, pure, unrefined, and nutrient-rich sweetener. Does reality match this perception?

We are indebted to Native North Americans for their introduction of maple syrup into our diet. They used a hatchet technique called "boxing" to tap maples by making either a diamond or Y-shaped incision into the tree. At the bottom of the cut they drove the hatchet deeply into the base of the "box" and inserted a flat splinter to guide the dripping sap into a wooden trough. They concentrated the sap either by boiling it in earthenware pots or by freezing it repeatedly and discarding the sap ice. Early settlers followed similar practices, which were recorded in family Bibles. Maple syrup and maple sugar were valued, since sugar cane was scarce and costly. The maple products varied greatly. Some iron boiling kettles produced syrups of light color and mild flavor; others, dark and strong. If the syrup was allowed to boil further, a dark mass of burned crystallized sugar formed. In 1794, one Vermonter described maple syrup production as one made under so many unfavorable circumstances that "it appears for the most part rough, coarse, dirty, and frequently burnt, smoky, or greasy."

By the 1850s, maple syrup processing equipment was transformed radically. Stave buckets replaced hollowed-log troughs. Horse or ox-drawn sleighs replaced sap yolks, and sap gathering in the sugarbush became easier. Syrup quality was controlled better when sheet-iron boilers with wooden sidepieces replaced iron kettles. The boiling equipment was suitable for indoor housing and the sugarhouse evolved.

As the population increased and land availability decreased, maple syrup producers tapped the same trees yearly. They learned conservation and used drill bits only large enough for reed spouts, called spiles, which inflicted less damage to the trees.

Even after cane sugar became more accessible and relatively inexpensive, maple syrup continued to be popular due to its distinctive flavor. Maple syrup reached markets farther and farther away from its origin. In 1898, Vermont syrup products were exhibited at the Philadelphia Exposition. The demand for maple sugar exceeded that for sugar. By 1890, Vermont maple production had peaked at about 14 million pounds yearly, but by the turn of the century, had declined.[39] To bolster the market, the University of Vermont and the State Agricultural Experiment Station jointly conducted research on maple syrup production, and many changes resulted from these efforts.[40]

Current maple syrup production still consists basically of tapping trees, gathering, concentrating and filtering the sap, and marketing the products. While modest cottage-industry efforts remain simple, large-scale producers use sophisticated equipment and techniques.

In 1962, FDA sanctioned use of paraformaldehyde pellets as sanitizers for maple tree tapholes. The pellets kill harmless bacteria normally found in maple trees, which retard sap flow. The long-lasting and slow-dissolving pellets keep tapholes open

longer and thus increase sap flow and profits. With pellet use
it became possible to begin tapping in the Northeast in January. Experiments showed that trees could be tapped even as
early as November, and then, by use of pellets, tapholes were
still productive the following April. The pellets have a lower
boiling point than syrup and presumably they evaporate during
processing. For safety, some producers discard the first sap
run.[41]

The pellet, sanctioned by Federal and state regulatory agencies, was denounced by some maple syrup producers. They
felt that maple syrup symbolized purity as a food and this
image would be lost.[42] With a sizable maple sugar industry
involved, the Canadian Food and Drug Directorate (now
Health Protection Branch, Health and Welfare Canada) re-
sisted pressure to sanction pellet use. Some American producers, having used the pellet, reported unfavorable results,
including tree damage below tap holes and early runs yielding
syrup with a leafy taste.

Pellet use may lower syrup quality. If the sap is collected
when the tree chemistry changes from "sweet water" to substances used for tree growth, the sap becomes "buddy" and
useless for producing a table-grade syrup. Pellet use increases
the possibility of producing buddy syrup. Even one or two
buddy-stage trees can contaminate and downgrade the entire
pooled sap from a sugarbush. When this happens, a bacterial
culture (*Pseudomonas geniculata*) may be added and allowed
to ferment, to mask buddy syrup and convert it into an acceptable commercial product.[43]

Maple syrup producers are not required to state on labels
that their products are from sap from pelletized trees. Some
do add label statements that their products are untreated, while
others, when questioned, may supply the information.

It was thought that the galvanized buckets and tin evaporators used in the past might leach minute lead quantities
into the sap. However, the sap did not remain in the collecting

buckets very long, and the evaporators quickly "limed" over with mineral deposits from the sap.

Current equipment used by most syrup producers differs significantly. While stainless-steel evaporators have soldered seams, most of them are blind-soldered, so that the boiling sap does not touch the soldered part. In large-scale operations, plastic bags may replace the metal collecting pails. Plastic pipe systems, permanently installed in the sugarbush, facilitate work in sap collecting. The pipes are cleaned by chemical flushing before the first sap run. While migration of substances leached from the plastics and sanitizing chemicals may occur, after long boiling such contaminants are not apt to be present.

Maple syrup grading is confusing for consumers, and as with many agricultural commodities, is based on appearance rather than taste. Color is the prime factor in grading, provided that the syrup meets requirements for density, flavor, and clarity. "Fancy" is the highest quality syrup by Vermont state standards, followed by Grades A, B, and C. But Federal USDA grades are U.S. Grade AA for Light Amber; A, Medium Amber; B, Dark Amber; and Unclassified. To many people, the highest Federal grade, AA, is the wrong color, being almost as clear as water. It is made from the first sap flow and lacks a strong maple flavor. Grade A, made from late season run, is darker and has a more distinctive maple-like flavor. Generally, the supermarkets stock Grade A. Grades A, B, and Unclassified account for 98 percent of the maple syrup market.

Currently, USDA is attempting to unify various Federal grading systems, for syrups, meats, and other agricultural commodities. The agency has proposed standardizing grades by use of letters A, B, and C. Under this proposal, Light, Medium, and Dark Amber would be grouped under U.S. Grade A for maple syrup, followed by the specific color. Such a revision, if enacted, would result in mayhem. Vermont Grade C would be upgraded to U.S. Grade B. To compound the confusion, imported Canadian syrup is graded both by number and color. Canadian syrup labeled No. 1 Medium would be equivalent to USDA Grade A Dark Amber, or to Vermont State Maple Grade B. Under the proposed revisions, the states may retain their own grades if they surpass Federal standards.[44]

Many Vermonters prefer their Grade B, and rarely is this grade exported outside the state. Grade B is less refined than Grade A, and aficionados claim that it has superior flavor.[45]

Through the years, maple sugar production has dwindled gradually. Around 1900, more than five million gallons were produced annually, but by 1972 production had declined to only one million. At the same time, demand increased. In 1971, market studies showed that some 18 million American households used pure maple products, mainly syrup, and some maple sugar. The use was infrequent. Unfortunately, however, the studies revealed that half of the consumers who *thought* that they were buying pure maple syrup actually had chosen blends with less expensive ingredients such as corn syrup, water, sugar, and artificial colors and flavors.[46] Such blends are known as pancake or waffle syrups, and are not new.

As early as the 1880s, one of the well-known blends was introduced in the American marketplace as an economical substitute for pure maple syrup. In the beginning, the product contained 45 percent maple syrup. As maple syrup became more expensive, over the years, the percentage of maple syrup gradually was reduced in the blend. Another well-known blend, introduced as recently as 1966, contained 15 percent maple syrup. But the percentage of maple syrup was reduced gradually in this product, too. Ultimately the level was so low that the company officials felt that the amount scarcely flavored the product, and decided to eliminate it totally.[47] A company representative observed,

> Americans have come to like — even prefer — the taste and consistency of imitation syrup. Our consumer tests showed that many people like the artificial stuff better than the real thing . . . Over the years, they've gotten used to it.[48]

This conclusion was reinforced by another survey, sponsored by the International Maple Syrup Institute. Of more than 2000 consumers, fewer than 200 could distinguish between pure maple syrup and imitations.[49]

Currently, one product labeled Canadian Brand Imitation Maple Butter is sold, with a prominent logo of a maple leaf. The product is made in Massachusetts and consists of glucose, cane sugar, milk, cornstarch, salt, and imitation maple flavoring. Another product labeled Vermont Maple Orchards Creamy Sugar and Butter Spread, contains neither maple syrup nor butter, and consists of 85 percent cane sugar syrup.[50]

Such practices have become so widespread that countermeasures have been started, jointly in a two-nation campaign, initiated by the International Maple Syrup Institute in Quebec and the United States Maple Syrup Information Bureau in New York. The effort is to protect consumers from misrepresentations through industry-wide use of a distinctive logo, showing sap dripping into a bucket, combined with a maple leaf. Display of the logo guarantees the purity and grade of the maple syrup. "We are advocating legislation which will assure the public that they are getting the food they are paying for," explained an Institute representative.[51]

Another countermeasure was initiated by the New York City Department of Consumer Affairs when a coffee shop was cited for violation by listing "real maple syrup" on the menu but serving a blend. Commissioner Elinor Guggenheimer reported, "Nothing in this syrup has ever been close to a maple tree. If maple syrup is listed on the menu, then it should be served — not a combination of chemicals, artificial flavorings, and colorings." [52]

From time to time, FDA seizes maple syrup products that are considered adulterated and misbranded, generally because cane or corn syrups have been substituted in whole or in part for maple syrup.[53]

In addition to the maple syrup blends, imitation maple flavors are available to food and beverage processors, in powder and liquid forms, for use in many processed foods including canned hams, processed meats, cereals, baked goods, candy, ice cream, and syrups.[54] One prominent flavoring company, advertising in a food trade journal, noted that its artificial maple flavor "tastes as authentic as a cold morning in Vermont." [55]

Is pure maple syrup a superior sweetener? When analyzed, pure maple syrup contained about 70 grams of total sugars per five tablespoons. This quantity, estimated as an average amount poured over a stack of three pancakes, served as a measure. Contrasted to this, the total sugars in syrup blends ranged from about 30 to 50 grams per serving.

In the same analysis, the predominant sugar in maple syrup was sucrose; in syrup blends, dextrose; and in one low-calorie syrup blend, fructose. Small amounts of fructose, maltose, and lactose were found in all the tested syrups. None of these sugars are more — or less — "natural" than any of the others.

Calories ranged from 272 to 313 per serving in all the tested syrups except the low-calorie one. That product contained about three times as much water as the other syrup blends, and had only 82 calories per serving.

The consumer testing organization that conducted this analysis suggested that, in many cases, choosing syrup blends would serve just as well as pure maple syrups and at one-third the cost. This advice may be thrifty but it's unwise. Syrup blends may contain fewer calories, but also fewer nutrients. In addition, they contain corn, cane, or beet fractions that may be allergens, while pure maple syrup may be tolerated.[56]

A Vermont newspaper headline proclaimed, "MANY NUTRIENTS ARE IN PURE MAPLE SYRUP." This statement was based on information supplied by Dr. Mariafranca Morselli, of the Botany Department, University of Vermont, who reported that maple syrup retains nutrients even though the sap is boiled, concentrated, and filtered. She claimed that pure filtered maple syrup contains "significant amounts" of calcium, potassium, manganese, magnesium, phosphorus, and iron; and trace amounts of riboflavin, pantothenic acid, pyridoxine, niacin, biotin, folic acid, and amino acids.[57]

While the nutrients Morselli listed so impressively may be found in pure maple syrup, they are not present in sufficient amounts to make any significant contribution to the diet unless the syrup is consumed at undesirably high levels. As with molasses and honey, maple syrup needs to be viewed as a pleasant flavoring agent rather than as a food that offers a dependable supply of nutrients. As with the other sticky syrups, it needs to be removed from the mouth quickly to avoid dental decay.

Sorghum

Sorghum, one of the oldest cultivated grains, yields sugar, syrup, and starch and also provides a human food crop and animal feed. Sorghum yields about 180 to 230 pounds of raw sugar per ton of stalk. Although sweet sorghum was recognized as a potential sugar source for more than a century, its full development awaited more recent research. The major sugar production methods used with cane and beet were ineffective for removing the large amount of starch present in sweet sorghum. While sorghum syrup could be produced, crystallized sugar could not. Conventional processes to recover sugar from cane or beet require temperatures near 200°F. At such heat, the sorghum starch granules gelatinize and thicken the syrup to reduce, or even totally prevent sugar crystallization.[58]

In the early 1970s, researchers at Agricultural Research Service, USDA, developed a technique to remove the sorghum starch by modifying sugar cane refining procedures. But the method was inapplicable to sugar beet facilities, and crystallized sorghum was not perfected for commercialization. By the mid-1970s, ARS announced that sweet sorghum soon might join the ranks of cane and beet as a source of crystalline sugar. In a factory production test, researchers were able to produce 22 tons of raw sugar from sweet sorghum equal to cane sugar's purity. This production was the culmination of a long, intensive effort involving the development of new sorghum varieties; improved growing, harvesting, and processing techniques; and innovative research. At last, it became possible to mill sweet sorghum in conventional mills and free the raw juice from starch by standard clarifiers. In contrast to other sweeteners, sorghum is minor, with an average annual American consumption of only one tenth of a pound per person.

Successful development of this new source of crystalline sugar was hailed as an aid to help America become less dependent on foreign sweetener sources.[59] An entirely different aspect deserves consideration. Is it ethical to divert a crop, traditionally used for human nourishment and animal feed, merely to satisfy our collective sweet tooth? [60]

Malt

To the general public, the word *malt* may be associated with malted milk or malt beer; to food and beverage processors, malt serves additional functions.

Bakers use diastatic malt, which is sprouted grain, usually barley. The grain is roasted with low heat to retain enzymes, then ground and made into either syrup or powder. Added to bakery dough, the enzymes in diastatic malt split the starches present in the flour into maltose and dextrin. These sugars, known as yeast foods in baking, assist fermentation and production of soluble proteins used by the yeast. These enzymes play comparable roles in the fermentation of beer and other malt beverages.

Bakers also use malt extract, which imparts good flavor and

color to baked goods. The extract is not very sweet, being at about the same level as corn syrup and only about a third that of invert syrup. Dehydrated malt extract is a readily handled crystalline form.[61]

Bakers also use malt syrup, which is dark colored and strong flavored, made entirely from malted barley. For lighter, sweeter syrups, malt extract is blended with corn syrup in various proportions and used by bakers as yeast foods. Higher percentages of corn syrup and lower percentages of malted barley are blended for use in confections, ice cream, vinegar, breakfast cereals, pretzels, puddings, flavorings, and pet foods, as well as in certain types of baked goods.

Traditionally, dextrins (polysaccharides made from cornstarch) were used as carriers of water-insoluble flavoring oils and other substances. After various processing steps, the dextrins helped to convert liquid flavors into dry powders, which then could be used in formulating many dry food and beverage mixes without flavor loss.

By legal definition, a food additive is any substance, other than a basic foodstuff, that is present in food as a result of any aspect of production. Hence, dextrins used for this purpose were regarded as a food additive. In 1958, dextrins were granted GRAS status. But more recently, with their greatly expanded use in processed foods, FDA curbed dextrins' use. Above established limits, dextrins are no longer GRAS.[62]

In recent years, maltodextrins were developed as replacers for dextrins as well as for gums and starches used in food and beverage processing. Maltodextrins are hydrolyzed carbohydrates. Some products are made from dent (field) corn, and the more expensive maltodextrins are made from more costly waxy maizes. Their sweetness level is lower than that of dextrose. As processing aids they serve many functions. Use of maltodextrins permits processors to spray-dry even the most difficult, highly hydroscopic food ingredients without clogging equipment. Dry mixes remain free-flowing without lumping or caking. Maltodextrins, unlike sucrose, have excellent suspension in solution and inhibit crystallization. Being bland, maltodextrins will not overpower delicate flavors in other ingredients. Because of their low reducing sugar content, maltodextrins can be used in baked products where corn syrup solids are undesirable. Maltodextrins are favored especially for creme fillings, glazes, snack food coatings, pie crusts, and cracker fillings.

Maltodextrins now replace dextrins to convert liquid flavors into dry powders and are found in many dehydrated products, including soup and gravy mixes; powdered citrus drinks; instant coffee, tea, and cocktail mixes; doughnut, cake, cookie, and icing mixes; spice blends, seasonings, and salad dressing mixes; and coffee whiteners. Additional uses include such candies as marshmallows and jelly beans; frozen eggs; frozen desserts; infant foods; peanut butter; whipped toppings; and sausages.[63]

Grain syrups

As an outgrowth of recent advanced malting techniques, wheat, rye, and rice syrups are produced and offer alternatives to malted barley syrups and blends. They give sweetness, enhance flavor and color, and add body and sheen to cereal-based products, such as baked goods, snacks, breakfast cereals, and pet foods. These syrups tolerate high baking temperatures, and each has its distinctive characteristics of flavor, color, and sweetness level. Other applications include beverages and confections. Use of these syrups may allow the reduction of other sugars and colors used in formulations.[64]

Rising sucrose prices have created interest in low-cost alternative sweeteners, especially when sweetening is not the ingredient's prime function. Among products that can replace sucrose is a by-product of beer brewing. Spent barley malt and rice grain, wastes from beer processing, contain soluble sugars that can be dried and milled to produce a sweet-flavored and colored sucrose replacer. The product can replace up to 20 percent of the sucrose in many food products, including breakfast cereals, breads, crackers, corn chips, and flavored snacks; up to 25 percent in cookies; and up to 35 percent in pancake and waffle mixes. The product, high in dietary fiber, provides texture, bulkiness, and water absorptiveness that helps keep products fresh.[65]

Whey

As Little Miss Muffet indulged in curds and whey, little did she know that whey was a potential sweetener. Spray-dried blends of sweet dairy whey, sodium caseinate, and lactose are available to processors who use such blends to replace up to 50 percent of sucrose in such products as creme fillings for snacks, cakes, and cookies.

For years, whey had been a dairy problem, for cheese-making produces large quantities of whey. Formerly, some of the whey was fed to animals, but much of it, viewed as waste, was dumped as effluent into streams. More recently, both with increased awareness of the need for pollution control and recognition of whey's nutritive value, the by-product has been utilized better. Lactose, the milk sugar in whey, became recognized as a potential sweetener; and by means of enzyme or acid hydrolysis of the milk sugar, lactose becomes more soluble and sweeter. However, hydrolyzed lactose still lacked stability and viscosity and tended to crystallize. Newer improved technology has improved these syrups, which are cheaper than corn syrups or sucrose and can replace traditional sweeteners to some extent in a variety of food applications, including ice cream, canned fruit, baked goods, beer, and confection. For example, hydrolyzed lactose syrups used in ice cream can replace up to 25 percent of the sucrose without altering the product's texture, "mouthfeel," flavor, or stability; in canned fruit, they can replace from 25 to 50 percent and maintain good appearance and flavor one year after processing; and in beer production, they can replace up to 50 percent of the malt extract and up to 20 percent of corn syrups without affecting product quality. In addition to their sweetening ability, hydrolyzed lactose products can replace the need for other food additives used to emulsify, stabilize, and form gels in many processed foods and beverages. Compared to sucrose, lactose has greater price stability. Although corn prices have been relatively stable, the possible future involvement in corn-derived gasohol may diminish corn syrup supplies. In this event, hydrolyzed lactose syrups may fill the void.[66]

*

All processed forms of sugars contribute approximately four calories per gram. All are primarily carbohydrates, and simple ones. Compare one cup of various processed sugars and syrups: [16]

type	grams	food energy (calories)	carbo- hydrates (in grams)
brown sugar, packed	220	821	212.1
molasses, 1st extraction, light	328	827	213.2
molasses, 2nd extraction, medium	328	761	196.8
molasses, 3rd extraction, blackstrap	328	699	180.4
molasses, Barbados	328	889	229.6
maple syrup	315	794	204.8
sorghum	330	848	224.4
table blend, mainly corn	328	951	246.0
table blend, cane and maple	315	794	204.8
table sugar, white	200	770	199.0

Obviously, all commonly added sugars are high in calories and do not differ dramatically one from another.

4

New Sweeteners: An Unjustified Craze?

Crystalline fructose, high-fructose corn syrups, aspartame, polydextrose

It is necessary to reinforce the idea that fructose is not a drug. It is only a natural food that in itself is deficient in vitamins, amino acids, and minerals. It is therefore a classic "empty calorie food." Yet it has advantages found in no other product.
— J. Daniel Palm, Ph.D., Department of Biology, St. Olaf College, "Fructose, the Empty Calorie Wonder Food," 1976

The advent of HFCS has made a basic and far-reaching change in the science of sweetening foods.
— Robert L. Martin, food industry engineer for ADM Corn Sweeteners, *Dairy and Ice Cream Field,* April 1979

The industrial sweetener industry has changed permanently and significantly in favor of corn products at the expense of sugar.
— J. William Leach, food and beverage analyst, Loeb, Rhoades, Hornblower and Co., *Business Week,* August 14, 1978

The future growth of HFCS . . . seems almost unlimited.
— *Food Engineering,* August 1979

Crystalline fructose (CF)

IN THE EARLY 1970S, new sweeteners were introduced, primarily through health food outlets, consisting of 99.9 percent pure crystalline fructose (CF). In a health food trade journal one CF product was described as "fruit sugar," which is "found naturally in honey, berries, fruits, etc.," and is "nearly twice as sweet [as sucrose] in cold foods and drinks; it looks and tastes like ordinary sugar but can save up to half the calories. It is safe for most diabetics and its use in place of ordinary sugar lowers the risk of dental caries." How accurate was this description?

While fructose is the most common natural sugar found in nearly all sweet fruits, berries, vegetables, and honey, CF is processed generally from sucrose, from cane or beet sugar.

Fructose's safety for most diabetics is *not* established, and its unrestricted use by diabetics could be dangerous. CF, like all sugars used in unrestricted amounts, may contribute to diabetes development.

CF is about one and a half times sweeter than sucrose in foods and beverages at room temperature, and up to two times sweeter in cold, acidic beverages; in hot liquids its sweetness increases as the liquids cool. CF has as many calories as sucrose.

Test results are equivocal regarding CF's ability to lower the dental caries risk.

The claim that CF as a sucrose replacer saves up to half the calories is theoretical only. In hot or cooked foods, there is no advantage. Also, the sweetness perception varies considerably, and to some individuals, CF does not taste sweeter than sucrose. The caloric reduction, if achieved, is minimal and minor.

What is this sugar? Fructose, also called levulose or fruit sugar, is a monosaccharide, and the sweetest of all commonly found natural sugars. Half of honey's dry substances consist of fructose. Sucrose consists of one molecule each of fructose and glucose.

In 1792, a German-Russian pharmacist, J. T. Lowitz, probably discovered fructose when he described a type of sugar that crystallized only with difficulty.[1] But fructose discovery usually is credited to a French chemist, A. P. Dubrunfaut, who in 1847 successfully isolated fructose from cane sugar. By 1874, the sugar achieved some practical significance when E. Külz, a German, suspected that fructose might be suitable for diabetics.[2] However, after Banting and Best discovered insulin in 1921, for three decades fructose's dietary use was nearly forgotten. During that time its use was limited almost exclusively to intravenous nutrition. It was produced by a complicated, expensive process, using inulin, a substance found in the tubers of such plants as dahlias, Jerusalem artichokes, and chicory.[3]

In the late 1960s, leading European crystalline fructose producers developed high-yield processes for large-scale economical production. Commercially pure CF generally is produced by separation of the fructose from the glucose molecule in sucrose by means of chromatographic selective absorption techniques. The fructose, which is four times sweeter than glucose, is crystallized to CF, and the glucose is sold as a by-product.[2]

By 1970, commercial quantities of CF were marketed in Europe, notably in Finland, West Germany, Sweden, and Norway.[4] In Finland, experimentally, more than a hundred commercially processed food items were fructose-sweetened to learn whether consumers would find such foods acceptable. Among the reformulated items were jams, jellies, marmalades, soft drinks, ice cream, frozen foods, baked goods, confections, pickles, mustard, catsup, and even marinated herring. All products were judged satisfactory, and many were rated superior to traditionally sweetened products. But even in Finland, which probably has the highest CF consumption in Europe, this sweetener is a minor sugar due to its expensiveness. Retail CF sales in Finland are only about 0.5 percent of all sugars sold.

Meanwhile, in the United States CF's potentials were largely unnoticed and underdeveloped by food and beverage processors. By 1975, CF began to attract attention. By then, an image for CF had been established as a "healthier" natural sweetening alternative for many general food and beverage applications, as well as potential applications for calorie-reduced foods. The special physical, chemical, and metabolic properties of CF were being extolled to processors, who began to view the sweetener as a particularly versatile, attractive raw material.[2]

CF is the most water-soluble of all sugars and dissolves readily, even in cool media. CF is very hydroscopic, which makes it an excellent agent to prevent baked goods and confections from drying out. Used in highly sweetened foods and beverages, CF does not crystallize out of solution during shipment and storage of products. CF is one of the most chemically reactive of sugars so it readily forms aromatically pleasant combinations and browning reactions in baked goods. CF, an excellent masking agent for the bitter aftertaste of saccharin, also increases its sweetness through synergism.[3] For example, 1.0 percent saccharin combined with 99 percent CF is five to seven times sweeter than sucrose. CF is synergistic with sucrose, too. A 10 percent water solution of a mixture composed of 60 percent CF and 40 percent sucrose is about 1.3 times sweeter than a pure sucrose solution, and 1.1 times sweeter than a pure CF solution, with both in 10 percent water solutions. By 1975, a food trade journal described the potential uses for CF as "limited only by the creativeness of food technologists and dieticians." [2]

Cost was CF's main drawback. Its refining process is infinitely more complex and expensive than that of sucrose. CF must be about 99.5 percent pure in order to make crystallization occur. No CF was produced in the United States, and Hoffmann–La Roche became the sole importer of Finnish CF. But soaring sucrose costs by the mid-1970s made CF somewhat more competitive. The fructose craze began, and gained momentum as many joined the bandwagon.

Increasingly, CF was used with many processed foods, including gelatin and frozen desserts, cake and cookie mixes,

and puddings; jellies, jams, and preserves; candies and chewing gums; salad dressings and mayonnaise; peanut butter;
protein supplements and other beverage powders; lemonade
tea mixes; isotonic beverages for athletes; and breakfast cereals
and their coatings for pre-sweetened cereals. One prominent
producer of frozen cakes, after successful test marketing,
launched national distribution of six new types of cakes sweetened with CF.

By 1979, fructose was a household word due to books and
media popularization. CF was central for diet plans purported
to quell cravings for sweets, appease the appetite, and lose
weight without dieting stresses such as abnormal hunger or
energy loss. "The Fabulous 14-Day Fructose Diet Plan" appeared in one popular homemakers' magazine.[5] A fructose
cookbook was advertised with hyperbole about the "sensational" sweetener, which was "all natural," "ideal for cooking
and baking, a natural preservative, an insulin stabilizer, and
a hunger reducer!" Food editors were educated about fructose's
advantages at a seminar sponsored by Hoffmann–La Roche
at the 1979 Food Editors' Conference.[6] One food editor reported that, in general, the audience remained unconvinced
that CF, at least ten times more costly at the retail level than
sucrose, was particularly beneficial for healthy people. At the
time, a survey in the Washington, D.C., area showed that CF
was selling at from $3.70 to $4.44 a pound; in packets, $6.08
to $7.04; in tablets, equivalent to a pound, $6.08 to $7.04;
and as liquid, $1.49, but the liquid consisted only partly of
fructose, and on a dry weight basis, would cost from $3.85 to
$4.71 a pound.[7]

Health food stores, health food sections in supermarkets,
and drugstores were targeted for marketing CF, as well as a
wide range of CF-sweetened products. One health food trade
magazine carried the following ad:

> Yes, we have fructose, the new sweetener you've been reading
> about. Fructose is an excellent source of energy for active

persons such as athletes, long-distance drivers, and anyone subject to stress and fatigue . . . Our pure unflavored fructose may be your answer to a healthier diet.[8]

In addition, liquid fructose, formerly available only to industrial and institutional users, reached the retail level, packaged in easy-pour bottles, intended as a general table sweetener and for baking, canning, preserving, and other home uses.[9] Honey-flavored syrup, containing 90 percent CF, was offered as an inexpensive honey replacer that would not crystallize or vary seasonally in composition, color, or flavor. One product, a yogurt-cake mix, contained dried yogurt and CF among its ingredients, simultaneously exploiting two "natural" food images. The marketplace was flooded with many CF-sweetened products, including "all natural" soft drinks.[10]

By 1979, the fructose craze reached such dizzy heights that Danny Wells, head of Standards Committee, National Nutritional Food Association (NNFA), a trade organization of the health food industry, sent its members a letter cautioning against misrepresentation of fructose and warning them not to place CF on a "majestic plateau." Wells warned especially against irresponsible statements that described CF as a "harmless sugar," "an excellent sugar for hypoglycemics," or worse, that claimed "even diabetics can safely use fructose." Wells suggested that retailers shun the fructose fad, and build sales based on sound nutritional principles.[11] Wells's justifiable warnings were given serious attention by responsible NNFA leadership, and at the organization's annual convention in July 1979 a program listed "The Fructose Debate" and provided a means to educate members.[12]

Within the health food movement, individuals engaged in soul-searching and questioned whether CF honestly could be regarded as a "natural" sugar, or rather be viewed as a processed carbohydrate. Shouldn't commercial fructose be considered a refined sugar, derived from another refined sugar, they asked.[8]

Food processors attempted to sell the natural image. Their persuasive argument was that, "as a result of its natural occurrence in many sweet tasting foods, fructose has always been part of the human diet." True, but incomplete. While commercial CF may be chemically identical to fructose, as found, say, in an apple or honey, the commercial product is an isolated substance; in food, it is accompanied by nutrients that aid in its metabolism.

The American Diabetes Association (ADA) issued a clear statement on this issue, of interest to both non-diabetics and diabetics:

> Fructose in its refined crystalline state contains few minerals, no vitamins, or fiber in contrast to naturally occurring fructose, e.g., fruit and complex carbohydrates that do contain these substances. The effects of ingestion of large amounts of any refined carbohydrate on this aspect of nutrition must be considered.[3]

In May 1980, the Center for Science in the Public Interest (CSPI) petitioned FDA and FTC to halt misstatements and exaggerations about CF's properties in advertisements and on labels. CSPI argued, in part, that manufacturers were taking advantage of "the sinking reputation of ordinary sugar" to promote CF-sweetened products as more healthful.[13]

Both agencies studied the question. The issue was difficult since neither agency had ever established a definition for a "natural" food; neither had the health food industry. In July 1980, in response to CSPI's petition, FTC announced that it had begun an investigation and expected that several of CSPI's concerns would be met by the agency's planned trade regulation rule on food advertising, which would cover food products' energy and low-calorie claims. FDA promised to have its field officers check labels.[14]

In late 1980, FDA received another complaint, this time from The Sugar Association, Inc., charging that certain claims made for CF-sweetened products were misleading. The association challenged the claims that CF is sweeter than sucrose and that less is needed to obtain the same sweetness level in a given food. In a letter to FDA, the association stated,

> It is difficult to make a sweeping generalization regarding the amount necessary to make a product acceptable to the consumer . . . [and] it has not been shown that because one caloric sweetener is sweeter than another its preferential use will result in a lowered intake of calories.[15]

In response, FDA agreed. John M. Taylor, director of the Division of Regulatory Guidance, FDA, reported that a food processor using CF in a product would mislead the public with claims that the "greater sweetness of fructose will result in lowered intake of calories because of a lesser use level compared to other sweeteners." [15]

Are claims justified that describe fructose as an "elite sweet," "healthier sugar," or "the miracle sweetener"? Hardly.

Metabolically, CF is somewhat different from other sugars. During digestion in a normal healthy individual, sucrose is broken down into equal parts of glucose and fructose. These components enter the blood stream through the small intestinal wall and the sugars are carried by the blood to the tissues and the liver. There, they are used or converted into glycogen, which is stored until needed by the body. Insulin makes it possible for glucose (blood sugar) to enter nearly all of the body's cells and be utilized. Both glucose and sucrose are absorbed rapidly, resulting in surges of high levels of blood glucose, which places uneven stressful demands on the insulin secretion system.[4]

Fructose is absorbed more slowly, and goes directly to the liver without insulin. The liver cells transform most of the fructose into glucose, which *then does require insulin* for use by the body's cells. Therefore, *it is incorrect to say that insulin is not required in fructose metabolism.* Almost all the fructose eventually becomes glucose, but it trickles into the system, and avoids large surges.

The low amount of insulin required in fructose metabolism and its slow absorption rate into the blood stream provide the bases for special beneficial claims.

One purported benefit is CF's value in weight-reduction diets. Fructose is less likely than sucrose to cause blood sugar level fluctuations, which may trigger low blood sugar (hypoglycemia) with its accompanying tendency to overeat and result in obesity. In 1979, the U.S. Postal Service moved against mail order diet books whose advertisements promised fast, automatic weight loss through CF use.[13]

Another claimed benefit is for persons undergoing extended stress periods, such as drivers, pilots, athletes, or individuals engaged in food fasts. The reasoning is that fructose has a protein-sparing effect due to its slow absorption from the gut but rapid metabolism in the liver. During exercise or fasting periods, the liver's glucose and glycogen reserves are depleted. In attempting to replenish blood sugar, the body may use protein as the necessary raw material. Fructose consumption can provide a long lasting supply of a readily available substance and thus spare protein from being used for this purpose. Fructose tablets are sold to athletes with the claim that by slowing down insulin release, fructose allows adrenalin to flow.

The most important claim, however, is fructose's special benefit for diabetics, who number some 10 million Americans. The slower uptake and the moderate blood glucose response are features that led to the statement that fructose is a better and easier sugar for diabetics to digest since they cannot manufacture enough insulin within their bodies to utilize glucose properly. These claims were reinforced by reports that fructose is widely used by diabetics in some European countries.[4]

Most fructose studies and experiments have been short-term or lack necessary safety and effectiveness data. Scientific studies to evaluate CF for long-term dietary diabetes management are non-existent. Studies with animals and humans have focused on tests with fructose and other non-glucose sweeteners in their pure forms, and as the sole item in the diet. According to the American Diabetes Association long-term realistic studies correlating fructose consumption with mixed meals typically eaten by diabetics are needed before any recommendations can be made.[16] ADA noted that the amount of fructose acceptable in a diabetic food plan is "uncertain at present," and in the absence of sufficient evidence, approval of CF for diabetic use can be neither accepted nor rejected.[17]

In well-managed adult diabetes, CF does *not* appear to have any dietary advantages over sucrose. In 1976, studies of blood sugar levels and urinary glucose excretion did not differ in nine adult diabetics compared with three normal controls, when all subjects consumed breakfasts with similar amounts of sucrose or fructose. The researcher concluded that *it seems unnecessary to have specially sweetened foods designed for diabetics.*[4]

Profound differences exist between the two main types of diabetes, and the effects of substituting CF for other carbohydrates in the diet also differ.[16] For example, obese diabetics may not be dependent on insulin, while lean youth-onset diabetics usually are. The advantages of using CF appear slight or negligible. CF use may be safe, provided the diabetic controls the total calorie intake.

Another consideration that tends to dampen enthusiasm for fructose substitution of sucrose was a demonstration in 1973 that carbohydrate restriction may not be as important in diabetic diets as previously believed. The new emphasis in diabetic management is to control the total caloric intake to achieve ideal body weight. This means proportionately restricting all foods, not necessarily disproportionately restricting carbohydrates.[18]

A report, prepared in 1976 for FDA by the Federation of American Societies for Experimental Biology (FASEB) indicated that most American diabetologists were *not* recommending CF use to patients. FASEB concluded, "It is the prevailing medical opinion that there are no clinical advantages of substituting fructose for glucose either orally or parenterally [intravenously] in any disease state."[4]

Dr. Victor Fratelli, assistant to the associate director for Nutrition and Food Sciences, FDA, reported that the agency regards the difference between the metabolism of fructose and glucose insufficiently significant to classify CF as a substance apart from sucrose. Fructose, which occurs naturally, was granted GRAS status for food use and was permitted whenever safe and suitable nutritive sweeteners are indicated. Because of this insufficiently significant difference between CF and other sugars, its value over other sugars for either diabetics or non-diabetics is questionable. Fratelli continued, "Any argument that one tries to develop that indicates these nutritive sweeteners are beneficial in weight reduction is very tenuous. What you are doing is substituting one carbohydrate for another."[4]

Diabetics were given commonsense advice by two diabetic authors, June Biermann and Barbara Toohey, in *The Diabetic's Total Health Book:*

> Fructose *cannot* be eaten freely without being counted. Indeed, one tablespoon of fructose is a fruit exchange — a fruit exchange without the fiber and nutritive advantages of fruit. A tablespoon of fructose does not seem like a good trade for a fresh, juicy peach or ten large, sweet cherries, or a small crunchy apple. Fructose, in short, is not the free lunch you've been looking for. What is a diabetic to do, then, to satisfy the sweet tooth that some diabetics seem to have? Rather than trying to satisfy it, we think you should just yank the rotten rascal out by the roots. Get yourself to the point that you no longer like excessively sweet things. Change your taste.[19]

Good advice for the general population as well as for diabetics.

Although CF's safety for hypoglycemics or its effectiveness for weight reduction are not proved yet, these features are used as selling points for many food products. A national natural-food, fast-food chain sells a popular fructose-sweetened "100 percent All-Natural Ice Cream" that the company claims has been designed specifically for hypoglycemics and diabetics. Another national distributor sells fructose packets and tablets as dietary supplements for appetite control.

ADA noted that in considering weight loss for the obese diabetic patient, fructose has the same number of calories as sucrose. "In the diet plan of the obese diabetic patient, calorie restriction, and not sugar restriction per se, with attainment of ideal body weight remains the major goal." [16]

Studies at Utah State University, reported in 1980, showed that calorie-counting individuals are not likely to benefit by substituting CF for sucrose. The researchers concluded that the only guaranteed way of decreasing calorie intake is to eat less (and exercise more). The researchers found, too, that contrary to popular belief and advertisements, CF does not sweeten all products more efficiently than sucrose, and therefore is undependable for calorie reduction. [20]

As early as 1972, ADA cautioned diabetics that

> many of the abnormalities produced by feeding sucrose (including increased cholesterol, triglyceride, enlarged liver, damaged kidneys, impaired glucose tolerance, insulin insensitivity) are also produced, but even more extensively, by the fructose (and not by the glucose) part of the sucrose. The harmful part of the sucrose molecule is the fructose half. The safety of regular consumption of large amounts of fructose is unknown. Accumulation of fructose (and sorbitol) in nervous tissues of untreated diabetics has been blamed for nerve damage. [21]

In rat studies, fructose raised the serum cholesterol level, while glucose showed no effect. Among potential adverse metabolic effects of high fructose consumption are its ability to induce gout in susceptible individuals and to raise fat levels (triglyceride and cholesterol) in the blood. Data are inconclusive, but many scientists and physicians consider high triglyceride levels possibly as a greater health threat than cholesterol.[22]

Metabolic disturbances may result in both normal and diabetic individuals when fructose is fed intravenously, either when infused rapidly or when used at high levels. Rapid infusions have caused nausea and epigastric pain. High infusion levels include a considerable rise in blood lactate and pyruvate levels, and metabolic acidosis may result, especially in uncontrolled diabetics. ADA advised that "considering the potential adverse health effects, administration of [intravenous] fructose should be discouraged, particularly in diabetic patients." [23]

Researchers suggested in 1974 that while the use of fructose offers no advantage in intravenous feeding, physicians mistakenly may believe that no harm is done because fructose is easily utilized without extra insulin.[4]

At times, physicians use fructose intravenously to sober drunken patients quickly, but studies have questioned the effectiveness of this use. Twenty males, aged 18 to 70, were hospitalized for treatment of acute alcohol intoxication. In double blind studies, the men were divided randomly into two groups of 10. Blood samples were drawn from all men. Then, each man in one group was given 100 grams of fructose in 1000 milliliters (ml) of water intravenously over a one-hour period; in the other group, a similar amount of glucose was given in the same manner. Two hours later, blood samples were drawn again. The blood alcohol levels for the two groups did not differ significantly, nor did members of either group differ in their ability to walk straight lines. However, fructose did induce metabolic disturbances, by raising uric acid and lactate serum levels in all 10 men. Reports from elsewhere have noted gout attack and lactic acidosis after fructose administration. The researchers recommended that fructose not be used to treat acute alcohol intoxication.[4]

Other metabolic disturbances have been reported when fructose is consumed at high levels. Diabetics promptly spill sugar into the urine. Hypoglycemics, too, can have adverse reactions. Large single oral fructose doses (70 to 100 grams) by normal healthy people may result in gastrointestinal upset, including diarrhea, flatulence, and colic pains. But large individual differences exist in human tolerance.[24]

Another health-related benefit ascribed to CF is a lower caries induction rate than sucrose. The evidence is equivocal. The claim is based on studies conducted in Finland from 1972 to 1974 at the Institute of Dentistry, University of Turku. CF totally replaced sucrose in the diet of human subjects. A four-day fructose-sweetened diet decreased plaque formation by 30 percent, compared to a similar sucrose-sweetened diet. However, while plaque is believed to be essential for caries production, visibly observed plaque is not necessarily cariogenic [leading to caries formation].

In two-year human feeding studies at Turku, 38 subjects were fed on a fructose-sweetened diet, and 52 on a sucrose-sweetened one. The dental caries incidence dropped by more than 25 percent in those fed fructose.[4] While results from this single study may be significant, fructose as a total sucrose replacer is quite improbable.

Another study, using realistic dietary patterns, showed that fructose equaled sucrose in contributing to decay on the crown's smooth surfaces in human teeth.[25]

Furthermore, fructose is a fermentable carbohydrate.[26] Fructose was reported in at least four different animal studies, from 1966 to 1974, to be approximately equal to sucrose as a suitable medium for bacterial growth and caries. However, possibly fructose may be more cariogenic in experimental animals than in humans.[4]

In 1977, *Streptococcus mutans,* the bacteria responsible for dental caries, was reported to be transported and metabolized efficiently by fructose. Thus, fructose can induce caries. As fructose is used in more foods and beverages in the American food supply, its advantage, if any, in lowering tooth decay, appears to be minimal.

Is fructose safe for everyone? Not for a small minority. Fructose intolerance, a rare inherited metabolic disease, is caused by the body's inability to handle the fruit sugar, regardless of whether it is present as a component of food or is in isolated CF form. Identified only as recently as 1956, the disorder is caused by lack of a liver enzyme. A fructose tolerance test is available, and once the condition is identified, treatment consists of a strict avoidance of fructose-containing foods. Undiscovered, fructose intolerance can be fatal to infants. Signs of the disease in infants include jaundice and enlarged liver, with vomiting, food rejection, and failure to thrive. In older children, enlarged liver also is symptomatic, as well as fits characterized by hypoglycemia. In adults, unrecognized fructose intolerance, characterized by tremors, sweating, vomiting, confusion, and convulsions, is apt to be dismissed as "neurotic ill-health." [4]

Consumption of CF probably will increase due to technological advances. The present method of processing CF uses sucrose as its raw materials. [27] In 1980, however, a new process described as "startling" would convert corn into crystalline fructose. According to *Business Week,* by using corn rather than sucrose as a base, fructose may "compete head on with cane and beet sugar." A U.S. pilot plant, utilizing the new process, was opened ahead of schedule in mid-1981. Predictably, future large-scale, efficient, less costly domestically produced fructose will increase the likelihood that the food supply will be laced with fructose, and many dubious products will be touted as "health" or "fitness" foods and beverages.

High levels of fructose consumption are undesirable for anyone, whether diabetic or non-diabetic. Fructose is neither a miracle sweetener nor a panacea. In *limited* quantity, fructose may be shown to be safe and effective for various applications, or merely for those who wish to shun artificial sweeteners or switch from other sugars. But consumers, tempted by an ever-increasing array of highly sweetened foods and beverages in the marketplace, find it increasingly difficult to limit total sugar consumption.

High-fructose corn syrups (HFCS)

"Fructose fest" luncheons were given to food editors and supermarket executives during 1980, with a menu consisting of "fructose punch, salad with fructose dressing, ham and fructose glaze, fructose yams, fructose green beans, honey-flavored fructose corn bread, and fructose sundae." The fructose insinuated into all items on the menu was not CF from sucrose but high-fructose corn syrups (HFCS) from corn. The National Corn Growers Association sponsored the luncheons as part of its public relations blitz during a lively competition between the corn and sugar interests, sparring to capture the sweetening market. By 1980, the public was confused about fructose. Many people thought that they were eating products sweetened with the much-touted CF when in reality they were eating products sweetened with HFCS. Even some reporters covering the Fructose Fest were confused and described the luncheon items as if they contained CF.[28]

For many years, corn refiners supplied the food industry with corn syrups and dried corn syrup solids. Their share of the sweetening market from 1950 to 1976 increased from 10 percent to nearly 25 percent. But technical problems limited corn sweetener use. Cornstarch, broken down into the glucose molecule, is only half as sweet as sucrose. Scientists searched for economical ways to convert corn sugar (dextrose) into the much sweeter fructose.[29]

In the early 1960s, Japanese researchers achieved the desired result. The Clinton Corn Processing Company obtained American licensing on the basic Japanese process and in 1964 began to perfect HFCS processing.[30] Three years later, Clinton introduced the first full-scale commercial American HFCS product.

The beginning stages of making HFCS are similar to the early stages of making conventional corn syrups. Cleaned corn is steeped in sulfurous acid to soften the kernels. Then, in a four-step process known as wet milling, the corn germ is separated and removed, the remaining fiber washed, the gluten separated and removed from the starch, and the remaining starch is washed. At this stage, the starch slurry is saccharified as far as possible by means of enzymes to make HFCS. The saccharified starch, which is liquid dextrose, is filtered, and through use of an ion-exchange treatment, is refined. Then the liquid is processed through an immobilized enzyme system so that less than half of the dextrose is converted into fructose in syrup form. By means of another ion-exchange treatment, the syrup is filtered and refined again. To obtain higher fructose levels, additional fractionation is done at the early separation steps.[31]

Clinton's first HFCS product contained a low level of 14 percent fructose. By 1968, Clinton's product contained 42 percent fructose, 52 percent dextrose, and 6 percent other sugars. The higher fructose level made the syrup sweeter, and the product was more competitive with sucrose. Later, second-generation HFCS products contained 55 percent fructose, 42 percent dextrose, and 3 percent other sugars; and third generation, 90 percent fructose, 9 percent dextrose, and 1 percent other sugars. These various rates gave processors flexibility for various food and beverage applications.[31]

HFCS has desirable attributes for use in food and beverages, being an exceptionally clear liquid, virtually water-white and with low viscosity.[30] While dry or granulated sweeteners are good for table use, liquid sweeteners are pre-ferred for industrial uses.[32] Processors of HFCS, hoping to invade the sucrose market, drew attention to HFCS's unique selling points in advertisements aimed at food and beverage processors.[33] HFCS's color is lower than that of liquid sugars from beet or cane. HFSC does not produce floc (fine sus-pended particles), which sucrose sometimes does. In acid foods, HFCS has no inversion, while sucrose can undergo chemical changes during processing that continue during storage and result in continual composition changes in per-ceived sweetness, flavor, and appearance. HFCS penetrates products such as fruits and sweet pickles faster than sucrose alone, and produces full-bodied, firm-textured products with satisfactory sweetness and color. HFCS has a synergistic effect on saccharin; used in combination, lower levels of the artificial sweetener can be used and the overall sweetness level is re-tained. HFCS can replace invert sugar completely in baked goods, and provides excellent browning effects. The clinching arguments were that HFCS is sweeter than sucrose and "of-fered very substantial economic savings." [34] In addition, since HFCS was not an artificial sweetener, processors believed that it "could help fend off consumer and health groups pressuring Americans to reduce their sweetener intake."[35]

By the mid-1970s, the traditional sucrose-dominated sweet-ening market was challenged by a combination of factors. Years of low prices and poor returns had slowed the inter-national expansion of sugar production. Worldwide sugar de-mands had outstripped production capacity. Everywhere, sugar stocks were exceedingly low due to greatly increased per capita sugar consumption. The supply dwindled further after an especially poor 1973–1974 sugar harvest in western Europe and the Soviet Union due to poor weather. The prob-lem was exacerbated by general worldwide inflation and by speculators who entered the sugar market. Also, many American farmers had switched from sugar beet production to more lucrative crops, such as soy, wheat, corn, and tomato, resulting in greater dependence on foreign sugar imports.

In January 1974, raw sugar sold on the American market for about 11 cents a pound. The price began an upward spiral and, by November 1974, reached an all-time high of 71 cents a pound. The precipitous rise caused major worldwide upheavals within the food industry. At this particularly critical time, HFCS was 32 cents a pound; dextrose, 21 cents; and conventional corn syrups, 12 cents.[36] Corn refiners were well-positioned to compete in the sweetening market with considerably lower-priced products. Wide-scale commercial production of HFCS made this particular corn-based sweetener a formidable challenger of the traditional cane- and beet-based sucrose.

In the sweetener market, sucrose symbolized instability; HFCS represented security. Corn supplies were ample and available. Corn, the world's most abundant food crop, also is a basic U.S. crop. The American corn refiners utilize only about six percent of the total domestic corn crop. Hence, corn sweeteners are not a major factor governing corn prices. Over the years, corn prices remained relatively stable. HFCS would protect the food industry against the vagaries of the world sugar market. Corn refiners claimed that they could produce enough corn sweeteners to replace all the sugar imported into the American market. The price differential between HFCS and sucrose was expected to increase in HFCS's favor, and this prediction was fulfilled. By 1980, its cost was at least 10 percent lower than that of sucrose.[37]

Sugar has few by-products to reduce its production cost. Some molasses and furfural (an aldehyde used to make phenolic resins) are minor by-products. Corn sweetener prices can be offset by up to 50 percent of the raw material costs by the sale of numerous profitable by-products.[38] Edible corn oil and corn germ are used in the human diet; premium protein ingredients from gluten seed and meal, as well as the soluble kernel portion, in livestock and poultry feeds; and starch, in various food and industrial applications. HFCS could be a partial replacer at various ratios, with no perceived difference in the finished food product nor changed consumer acceptance. In some cases, HFCS could replace sucrose totally, pound for pound on a dry weight basis. There were relatively few limitations. HFCS could not be used in dry desserts,[39] or in candies because it would make the product sticky, or in ice cream because it would cause a lowered freezing point. However, university-industry–sponsored research programs fine-tuned the applications, and HFCS even became a total sucrose replacer in ice cream manufacture.[40]

HFCS, containing fructose at 42, 55, and 90 percent levels, rapidly achieved success in the sweetening market. In 1967, HFCS's annual per capita consumption was 0.1 pounds; by 1980, 18.9 pounds. Currently, processors use HFCS in a wide range of products, including carbonated and non-carbonated beverages; jams, jellies, and preserves; salad dressings, pickles, and catsups; baked goods and icings; cereals; dairy products, such as chocolate milk, eggnogs, yogurts, ice creams, and sherbets; diabetic and reduced-calorie foods; table syrups and liquid table sweeteners; and at least one processor uses the sweetener in wine for attributes other than fermentation and alcoholic content.

A massive switch occurred in 1980, when the two major cola beverage companies announced plans to use HFCS in their products. Sweetener use by these two companies represented two million tons of sugar annually. This news sent other processors scurrying to place HFCS purchasing orders as a hedge against possible shortages, while HFCS producers retooled plants to increase production for anticipated sales to other soft drink manufacturers.[41] By the mid-1980s, the anticipated annual market for HFCS is expected to be eight million tons.[42]

As HFCS saturated the food supply, public confusion about fructose increased. The Sugar Association, in complaint to FDA, charged that several companies were making label claims that their products were made with fructose, although the products were sweetened with HFCS. By late 1979, FTC began to scrutinize various advertising claims made for fructose- and HFCS-sweetened products, prodded in part by a new petition from the Center for Science in the Public Interest that targeted on advertisements for a specific soft drink, Shasta. The advertising copy read that the drink was "an alternative to ordinary sugared soft drinks," and claimed that ordinary refined sugar was replaced in the product "with simple sweeteners like fructose, a kind found in fresh fruit." CSPI claimed that the drink was sweetened with 55 percent HFCS and the advertisement was deceptive on three counts: HFCS is not pure fructose; fructose is refined; and HFCS is not found in fresh fruits and honey.[43] While FTC was deliberating,[44] news broke elsewhere.

An inner office memo, dated January 25, 1980, read "Today Botsford Ketchum resigned the Shasta account." Botsford Ketchum, a prominent ad agency, had the six million dollar annual Shasta account. In a memo to Ketchum employees, the company's president stated,

> We believe in the principles we stand for and the people who have helped build this agency. We do not intend to change our course. Consequently we believe Shasta will be better served by a different kind of agency and that our people can work better on other accounts.

The principle involved Shasta's advertising statements concerning fructose. The story, pieced together by *The Washington Post,* demonstrated that advertising agencies may depend on clients for technical information about products for advertising copy. Botsford Ketchum appears to have relied on Shasta's supplied information about fructose. But eventually, through outside sources, the agency concluded that better information was needed. While the advertisements and television commercials had been cleared by both legal counsel and television network scrutiny, it was unlikely that either the agency's own lawyers or the networks' standards and practices departments were aware of the technicalities involved. After publicity of the case, the company changed the thrust of its advertisements and commercials and labeled Shasta as a product with a high-fructose corn sweetener.[45]

When HFCS-containing products are properly labeled, is the sweetener safe for everyone? Individuals with at least two types of identified health problems need to avoid HFCS: Corn-sensitive individuals must avoid this sweetener since it is corn derived, and diabetics need to avoid HFCS because of its glucose content. All HFCS contains glucose, with some products having as much as 58 percent. Glucose is the sugar that diabetics are *least* able to metabolize properly. In a 1980 press release, ADA noted public confusion and warned diabetics to exercise extreme caution when purchasing foods whose labels claim that the products have been sweetened with fructose, since the sweetener, in reality, may be HFCS.[46]

At a seminar in 1980, the food industry, too, was cautioned by Dr. Howard R. Roberts, Director of Scientific Affairs, National Soft Drink Association (and former Director of FDA's Bureau of Foods), that HFCS is not pure fructose but consists of glucose, too.

> Industry should not be lulled into a false sense of contentment by thinking that increased use of corn sweeteners will alleviate the oversimplified criticism of sugar. Whether fructose comes from ingested sucrose or is ingested as fructose per se, it ends up as glucose in the body and the body cannot tell where the glucose originally came from.[47]

Aspartame

The case of aspartame is a paradigm for the complexities involved for petitioners, regulators, and consumers, in attempts to launch a new alternative sweetener.

Aspartame is a white crystalline low-calorie sweetener synthesized from two amino acids — protein's building blocks — normally found in the human body. One, aspartic, is flat tasting; the other, phenylalanine, is bitter; yet combined, these two amino acids form a sweet tasting compound.[48]

This accidental discovery was made in a G. P. Searle Company laboratory in December 1965, by a scientist developing ulcer therapy drugs, who noted that one combination of amino acids had an especially clean, sweet taste. The compound was L-aspartyl-L-phenylalanine methyl ester; Searle named it aspartame. The substance left no unpleasant aftertaste. It was 200 times sweeter than sucrose and could intensify the taste of other flavors and sweeteners; aspartame made table sugar taste 23 percent sweeter, and saccharin, 30 percent sweeter. Recognizing its commercial potential, Searle in combination with a Japanese company, Ajinomoto, researched aspartame for nearly a decade, and the prospects of commercial production appeared promising.[49]

In July 1974, aspartame gained FDA approval for all uses except with cooked foods and bottled soft drinks, since heating causes aspartame to break down into diketopiperazine (DKP). Although this tasteless chemical was thought to be harmless, FDA requested further data. Searle agreed to submit additional studies. FDA viewed DKP as a technological problem, not a safety issue.[50]

A month after aspartame's approval, John W. Olney, M.D., charged that FDA "has presented a seriously misleading case for the safety of aspartame." Earlier, Olney had reported test results demonstrating that monosodium glutamate induced brain lesions in infant mice and rats and in one newborn monkey. Olney and his colleagues had studied both glutamate and aspartate (salts of two amino acids, glutamic and aspartic), as well as their structural analogs. Olney reported that both amino acids are similar in molecular structure; both are neurotoxic and capable of inducing brain lesions. Aspartate, fed to young mice, resulted in the same type of brain damage induced by glutamate.[51]

Information that was being given to food processors through trade journals had emphasized the idea that aspartame was not artificial, but composed of two amino acids. The fact that aspartame is made of naturally occurring substances is no reason to assume that the new combination is safe. FDA had already removed certain amino acids, used as nutrients, from the GRAS list and made them subject to regulated standards for purposes of food fortification. Amino acid use by the food industry is regulated, as it should be, since amino acids can produce toxicity as well as imbalances.

Olney was concerned that children especially were at risk, due to their high intake of sweetened foods (which might be flavored with aspartame) plus their high intake of processed foods containing the frequently used food additive monosodium glutamate (which contains glutamate). For this reason, according to Olney, FDA's margin of safety for aspartame fell far short of the traditional safety margin for adults and was mathematically off-base a hundredfold for children.[52]

Apart from the potential problem of the aspartate fraction in aspartame, the other amino acid, phenylalanine, held another threat. This amino acid is crucial in phenylketonuria (PKU), an inborn metabolic error which, if untreated, can lead to irreversible mental retardation. Damage can be avoided by early diagnosis and rigid adherence to a phenylalanine-free diet. One in 10,000 children is born with PKU, and Olney emphasized that appropriate warnings were necessary. Searle proposed a label statement on aspartame-sweetened products: PHENYLKETONURICS: CONTAINS PHENYL-ALANINE.

Olney, the attorney James Turner (author of *The Chemical Feast*), and a consumer group jointly made four charges. FDA had acted too hastily in granting aspartame approval. FDA's decision was based on industry research. FDA's studies were limited to taste tests, not safety. Consumers were excluded from FDA's review of Searle's data.[53]

Turner, representing Olney and others, petitioned FDA to allow public access to the data, withdraw the sweetener's approval, conduct more detailed studies, and hold public hearings. Senator William Proxmire (D.–Wis.) entered the fray, charging FDA's commissioner with "malfeasance in office" for having granted approval without hearings. Proxmire alleged that the investigation and approval processes were "rigged to protect the manufacturer."

On December 5, 1974, FDA stayed aspartame approval and forbade any further marketing until all safety issues could be aired and resolved. The agency scheduled a board of inquiry and announced its intention of using a trial judge.[54]

Unforeseen events broke that made it necessary to postpone the hearing. DKP, the breakdown product that formerly had been considered harmless, now was under scrutiny for its potential to combine with nitrite in the human stomach and form carcinogenic nitrosamines.[55] Other warning signals were flashing, too. DKP appeared to cause liver cancer in rats and uterine polyps in rats fed medium to high doses.

In December 1975, in an audit of Searle's animal study records, FDA noted discrepancies and questioned the test data reliability for aspartame as well as for three Searle drugs. FDA organized a special task force to evaluate the data. After review, the group concluded that Searle had been "careless in its conduct, evaluation, and reporting of animal studies."[56]

In April 1976, FDA requested a grand jury investigation of Searle's drug-testing practices. At the same time, FDA continued to prohibit aspartame's marketing until further evaluation of animal safety studies. Several unresolved safety questions remained: What were the long-range DKP effects?

Was DKP hazardous because of its potential to form nitro-samines? What was the significance of certain pathological findings, such as brain tumors and liver and kidney changes noted in some test animals in a lifetime study? What was the significance of adverse findings in a study of newborn rats? What was the meaning of an increased incidence of hyper-plasia (an abnormal increase in the cell numbers in a normal tissue arrangement) in aspartame-fed mice? and tumors ob-served in urinary bladders of DKP-fed mice? and in a 26-week study, using urinary bladder implant? To clarify these issues, Searle submitted additional data.[57]

FDA decided to use an outside advisory panel. The Uni-versities Associated for Research and Education in Pathology (UAREP) would review the data. While this organization was regarded as an independent scientific group, critics ques-tioned the panel's ability to be impartial. Searle had contracted for the review.[58]

By 1979, to determine the data's accuracy and reliability, UAREP had worked its way through only 12 out of some 100 Searle's studies. The panel found "no indication that the results of animals in any test group had been altered de-liberately to produce biased results."[59]

This validation of research techniques and records led once more to FDA's review of aspartame safety data. The agency decided to convene a scientific board of inquiry, a pilot project for the agency, and a departure from its usual practice of a formal evidentiary hearing before an adminis-trative law judge. The new format was chosen in order to provide a forum to discuss scientific issues and, it was hoped, to resolve them without having to resort to legal formalities.[60]

Despite the claimed advantages of the format, it was greeted warily.[61] It was untested and might prove to be in-appropriate. While technical controversies might benefit, scientific, legal, or political issues might be resolved better during evidentiary hearings. The new format lacked any pos-sibility for cross-examination of testimonies. In theory, board use might be time-efficient for making a decision, but in reality, a series of hearings could be requested that actually might prolong the procedure. Also, if a petitioner did not agree with the board decision, he needed to challenge it in court. In this case, he would be back to square one, with an evidentiary format.[62]

The board of inquiry consisted of three members: one selected by Olney and Turner, one by Searle, and one by FDA.[58] As the inquiry progressed, news from elsewhere seemed like promising omens. The Food and Agriculture Organization and the World Health Organization's Joint Expert Committee on Food Additives had approved aspartame for human consumption. The sweetener had been approved and was being sold elsewhere. Officials within FDA's Bureau of Foods had indicated that they favored approval.

During 1979 and early 1980, Searle and Ajinomoto, as well as food and beverage processors, were cliffhanging, awaiting the decision. A possible $1.5 billion annual market in sales of aspartame was involved, and a vast array of food and beverage applications.[62]

Olney's position was being strengthened by unresolved safety issues and by new ones that had developed since the 1974 staying action. Other scientists supported Olney, notably Floyd Bloom of the Salk Institute, who believed Olney's evidence demonstrated that even moderate amounts of aspartate are neurotoxic.

The board of inquiry conducted marathon sessions, with charges and countercharges. The issue of possible brain tumors generated the greatest controversy. The three-member board said it had "no choice but to conclude that the data reported ... do not rule out an oncogenic [tumor-causing] effect of aspartame," and that the findings "appear to suggest the possibility that aspartame, at least when administered in the huge quantities employed in these studies, may contribute to the development of brain tumors." [63]

At the end of the inquiry, the fate of aspartame remained

in limbo. The safety had not been proved with reasonable certainty, and the board recommended that FDA withhold approval until the questions could be answered satisfactorily. At the very time that this decision was being given, Searle was bringing suit against FDA in district court to force a final decision from the agency. The two actions came only hours apart. FDA denied any connection and reported that the proximity was purely accidental. Searle filed a lengthy scientific rebuttal to the board's conclusions, taking exception to interpretations made from the presented evidence. Searle offered new expert opinions on aspartame's safety.

Searle's action forced FDA to organize a new scientific board of inquiry. The board, comprised of three outside experts, held hearings during January and February of 1980 and by October of 1980 gave a qualified disapproval of aspartame. The board recommended that no approval be granted until further long-term animal tests could be conducted in order to rule out the possibility that aspartame might cause brain lesions or tumors. The board agreed with Olney that the cancer tests should be redone, since the question of irregularities was impossible to resolve. However, the board admitted that the evidence, to date, did not support the charge that aspartame might kill clusters of brain cells or cause other types of brain damage.

The ultimate decision to approve or disapprove aspartame use rested with Arthur Hull Hayes, Jr., M.D., the newly appointed commissioner of FDA. Hayes had to weigh the balance of a staggering load of data and opinions. On one side of the scale were the recommendations of two independent boards to withhold approval. On the other side of the scale was the position, already taken before Hayes's appointment, of the Bureau of Foods of his own agency, which favored approval. Meanwhile, France, Germany, Belgium, Switzerland, Denmark, Luxembourg, Brazil, the Philippines, Tunisia, and Mexico had joined a growing list of countries where aspartame use was permitted. In addition, after the convening of the last board, FDA received the results of a completed long-term feeding test, conducted for Ajinomoto supporting Searle's position.

Two additional factors may have helped tip the scales. A developing political climate favored the philosophy of "less government interference" and concept of "risk versus benefit," which pervaded FDA as well as other Federal agencies. Also present was the uncertainty of saccharin's ultimate fate. If, at some future date, saccharin were to be banned, there would be no alternative non-nutritive sweetener. Aspartame, as a low-calorie sweetener, to some extent might fill the void, although its present manufacture is far more costly than saccharin.

On July 15, 1981, Hayes caught the business world and food processors by surprise when he announced that FDA would approve aspartame for use by food manufacturers of cold cereals, drink mixes, sugarless gums, instant coffees and teas, gelatins, puddings and fillings, and dairy products and toppings. Aspartame would also be approved for retail sale in tablet form and as a free-flowing sugar for home use. Approval did not extend to certain other uses, since Searle had not sought approval for aspartame use in carbonated soft drinks or baked products. As formulated, aspartame loses sweetness with heat or long storage of products.

Two weeks later, the Canadian Department of Health and Welfare approved aspartame, including its use in carbonated soft drinks.

After the Canadian approval, Searle announced that it would seek approval for aspartame use in carbonated soft drinks for the United States market.

In approving aspartame, FDA established a new precedent. Searle, the aspartame manufacturer, is required to monitor consumption levels of the sweetener by the public. Doubtless, the history of uncontrolled proliferation of cyclamates and saccharin was responsible for this measure. Such surveillance may provide warning signals and, if necessary, restrictions, long before consumption rises to a level that becomes a potential threat to human health.

In announcing aspartame approval, Hayes said, "Few compounds have withstood such detailed testing and repeated close scrutiny. The process through which aspartame has gone should provide the public with additional confidence about its safety." Olney, however, reacted to the approval by stating that he still believed that the substance could be a hazard. He predicted that there could be further attempts to halt aspartame use.[64]

Polydextrose

Artificial sweeteners supply the sweetness of sugars, but they fail to provide the bulk. This is a special problem for such food products as frozen desserts, instant puddings, and hard candies.

Polydextrose, a bulking agent, provides this feature. Polydextrose consists of dextrose, sorbitol, and citric acid. It is less sweet than common sugars, but provides only one calorie per gram compared to higher caloric levels of common sugars, carbohydrates, and fats.

Pfizer, Inc., had researched polydextrose for a decade. In 1978 the company submitted a Food Additive Petition to FDA. In anticipation of imminent approval, Pfizer publicly introduced polydextrose on May 28, 1981.[65] On June 5, 1981, FDA approved polydextrose.

Since polydextrose contains sorbitol, the well-known laxative effect of sorbitol is present in polydextrose as well. When consumed, most of this bulking agent passes through the gastrointestinal tract unchanged and is excreted. Due to the large unabsorbed portion, when consumed at a high level, polydextrose creates an osmotic load in the lower intestines and is laxative. In a study with adults, it was found that consumption of from 50 to 130 grams of polydextrose a day, with an average of 70 grams, induced an effect only slightly less laxative than that of sorbitol.[66]

Pfizer researchers believe that polydextrose is well tolerated in amounts apt to be consumed in foods for which it is approved. Nevertheless, as a precaution, when a single serving of a food contains more than 15 grams of polydextrose, the label of a food must state: SENSITIVE INDIVIDUALS MAY EXPERIENCE A LAXATIVE EFFECT FROM EXCESSIVE CONSUMPTION OF THIS PRODUCT.[51]

Artificial Sweeteners:
Should Carcinogens Be Allowed?

Cyclamates and saccharin

Artificial sweeteners may be the nutritional disaster of our time.
— George V. Mann, Sc.D., M.D., *Postgraduate Medicine*, July 1977

There is no theoretical reason to believe ... that the effect of cyclamate is proportionately much less at lower doses; with many carcinogens, the observed rule is that lower doses simply take a longer time to take effect.
— Dr. Joshua Lederberg, Nobel research scientist, Stanford University School of Medicine, 1970

One might expect that if saccharin is implicated in human bladder cancers, indications of it would be expected to turn up around the year 2000. Saccharin use peaked during World War II when sugar was unavailable. Sixty years would be about the time cancer epidemiologists would expect the disease to develop after such high exposure, based on the effects on offspring of mothers who had been consuming the sweetener.
— Dr. David Rall, Director, National Institute of Environmental Health Sciences, 1977

THE QUEST FOR NON-NUTRITIVE SWEETENERS is alluring to food and beverage processors. Satisfactory non-nutritive sweeteners permit them to develop a wide range of highly profitable products, with something for everyone. Diabetics and weight-watchers can indulge in forbidden foods without suffering guilt pangs. Parents consent to having their children consume quantities of sweets, with the comforting thought that the foods will not rot the teeth. And the general public, with its collective sweet tooth, can continue to gorge on pastries, cookies, confections, soda pop, and other sweetish goods.

However alluring, this quest has been a disaster. To date, the non-nutritive sweeteners introduced into the marketplace have been found, after investigation, to be hazardous as well as ineffective. Their use does not help to reduce weight; on the contrary, non-nutritive sweeteners stimulate appetite. The underlying philosophy that encourages the use of non-nutritive sweeteners is faulty. Dependence on these substances, whether it be for special groups, children, or the general population, encourages the use of highly processed foods rather than fresh basic ones that do not require added sweeteners. But the problem of non-nutritive sweeteners is not solely a nutritional issue. It has become an economic and political one, as the alluring quest continues.

Cyclamates

A young researcher at the University of Illinois, while studying various compounds as possible fever-reducing drugs, isolated the barium salt of *N*-cyclohexylsulfamic acid. He was smoking while working with this substance, and, casually brushing off some tobacco shreds clinging to his lips, he noted that his finger tasted intensely sweet. The researcher knew that the barium salt of the acid was relatively toxic, so he prepared the sodium salt. The year was 1937; the researcher, Dr. Michael Sveda; and the sodium salt was to be called sodium cyclamate. Its potential uses to mask the taste of bitter pills and to sweeten were recognized. By 1942, Sveda and his superior, Dr. Ludwig F. Audrieth, jointly patented cyclamate and its salts.

In 1950, Abbott Laboratories launched the first commercial

production of sodium cyclamate, and the following year marketed calcium cyclamate. Du Pont, Monsanto, Pfizer, Pillsbury, and others soon followed.[1]

At first, cyclamates were used for special dietary purposes, mainly for diabetics and persons who needed to restrict their weight. Cyclamate-containing products carried the same legend as saccharin, already in use: "Contains [brand name inserted] as a non-nutritive, artificial sweetener which should be used only by persons who must restrict their intake of ordinary sweets." This statement, devised by FDA in 1941, was intended to allow the marketing of saccharin-sweetened foods to meet special needs, but at the same time served as warning to sugar-rationed wartime consumers that the food value of sugar was not present in such products.

In time, however, cyclamates, as well as saccharin, came into *general* use. No clear definition existed regarding what constituted a dietary food. Processors labeled artificially sweetened foods as dietary merely because the omission of sugar spared a few calories. The dietary tag was used by packers and retailers as a promotional gimmick. FDA proposed a new labeling regulation that, if enacted, would have limited the dietary claim to those foods or beverages with significantly reduced calories. But the agency was deluged by industry protests, and the proposed regulations remained in limbo for decades. Meanwhile, by sly editing, processors transformed the term "non-nutritive" into "no calories!" and the phrase "should be used only by" into "recommended for." The cautionary warnings had been converted into selling points.[2]

In 1955, FDA had requested a cyclamate review by the Food Protection Committee of the National Academy of Sciences (NAS). The committee judged that cyclamates' limited use was not hazardous in low daily intake for special dietary purposes. While not challenging cyclamates' safety, the committee noted that the available scientific data did not assure safe unrestricted use by children, pregnant women, and persons who suffered from lower bowel diseases or other conditions.[3] Despite this warning, FDA took no action to limit cyclamates' use.

Cyclamates, less intensely sweet than saccharin, were used to mask saccharin's bitter aftertaste and were combined with saccharin in many low-calorie soft drinks, canned fruits, baked goods, soups, bacon, puddings, salad dressings, toppings, syrups, chewing gums, and toothpastes, in addition to many table uses to sweeten coffee, tea, breakfast cereals, and other foods.

In 1960, disquieting news came from Japan, where a woman and infant were examined at the Department of Pediatrics, Iwate Medical Center. Both suffered from malnourishment, but in addition, the baby was malformed. The physicians ruled out malnutrition or radioactive exposure as likely causes of the malformation. In questioning, the woman mentioned that early in her pregnancy she had eaten dulcin, an artificial sweetener that was cheaper than sugar. Her consumption was estimated to be about 0.3 grams daily, and it was thought that dulcin probably was responsible for the infant's malformation.[4] Dulcin, about 250 times sweeter than sucrose, was synthesized in 1883 and had been used for more than 50 years in the United States and elsewhere until long overdue safety tests were performed. Dulcin was found to cause liver cancer in rats, and was banned in the United States in 1950, but its use in Japan was still permitted.

This case led Professor Ryoza Tanaka of Iwate Medical Center to study the effects of dulcin and other artificial sweeteners on the mouse fetus. Cyclamates affected the fetus at a lower level than thalidomide, the drug that had caused gross malformations in the human fetus. Tanaka's findings, extrapolated from mouse to human, had important implications. The data suggested that a pregnant woman, drinking as few as two bottles of artificially sweetened soft drinks daily, might reduce her chance of having a normal live birth by 50 percent.[5]

In 1962, FDA requested another cyclamate safety review, and again the Food Protection Committee recommended that cyclamate use be restricted to special dietary foods. The committee raised questions about cyclamate safety for broader use, and stated that "the priority of public welfare over all other considerations precludes, therefore, the uncontrolled distribution of foodstuffs containing cyclamates." The committee noted that cyclamates consumed at high levels exerted a laxative effect, more pronounced in children than in adults. Data were needed concerning the effects of cyclamates on the embryo, infant, or young child, whose tolerances are likely to be lower than adults'. Studies had been limited to adult males, and data were needed on the effects on women, too. Little was known about the way the normal body metabolized cyclamates, nor the body's ability to metabolize these sweeteners by persons who had chronic diseases, such as diabetes, or disorders of the kidney or large intestine. The committee cautioned that "it is not unlikely that the per capita consumption of cyclamates will exceed five grams if the potential usages of this material in foods and beverages is exploited." [6]

FDA took no action to limit cyclamate use. According to estimates, the five-gram limit easily could be exceeded by persons who ate a variety of artificially sweetened foods and drank several bottles of diet soft drinks daily,[7] a dietary pattern that was not uncommon. Cyclamates were being used increasingly by individuals and by processors in an increasing variety of foods and beverages, and at higher levels. Food and beverage processors found that they could replace six dollars' worth of sugar with about 64 cents' worth of cyclamates.[8] According to a 1963 industry survey, 31 percent of all households used artificial sweeteners; before long, it would reach 75 percent.

The Food Protection Committee's reviews consisted of evaluating reports, not engaging in any research. The reviews were notably lacking in any sense of urgency. Critics noted that the committee, dominated by individuals with limited toxicological knowledge, tended to ignore possible subtle, long-term hazards. Furthermore, the group was either unable or unwilling to prod FDA into action to curb cyclamate use.[9]

Meanwhile, the medical community began to express concern over the uncontrolled and indiscriminate use of cyclamates, which they viewed as drugs. In 1964, *The Medical Letter* recommended a toxicologic reappraisal of artificial sweeteners in healthy persons, as well as in pregnant women and in mother's milk. Artificial sweeteners consumed by pregnant women are carried through the placenta into the developing embryo. Also, they are transported from a lactating woman into breast milk fed to the infant.

The Medical Letter, which did not cite the dulcin case, noted that while there was no evidence of harmful effects from cyclamates, data were incomplete. No research studies on artificial sweeteners had been published in recent years, and many unanswered questions about their toxicity needed to be resolved. Meanwhile, the current promotion and extended cyclamate use "are regarded as against the public interest, and action by the medical profession, health authorities, and Federal agencies to inform and protect consumers is in order."

The Medical Letter was critical of the claims for artificial sweeteners' effectiveness for weight control and deemed them "of questionable value." *The Journal of the American Dietetics Association* had reported no significant differences between weight losses of users and non-users of artificial sweeteners.[10]

Abbott answered charges raised by the medical publication and claimed that the safety research had already been conducted and results filed with FDA. *The Medical Letter* then claimed that unpublished evidence was inadmissible because the confidential data had not been reviewed by independent scientists.

The data were released. One study with laboratory animals purportedly demonstrated cyclamate safety by the absence of adverse effects on fertility, pregnancy, the fetus, and offspring. Another study, of only two weeks' duration, was offered as proof of absence of adverse effects in patients with impaired kidney function. Another study presumably showed that a large number of children showed no ill effects attributable to cyclamate consumption, but at the time, the study was still being analyzed. Other studies included one that concluded that the substitution of cyclamates for sucrose was weight reducing, and another that cyclamates passed through the body without pharmacologic effects. The last conclusion, asserted *The Medical Letter,* was refuted by Abbott's own data.[11]

FDA reviewed the studies as well as new experimental data. In May 1965, the agency concluded that there was no evidence that cyclamates, at present uses, were hazardous to health.[12]

Shortly after this announcement, several reports from different sources questioned both the effectiveness and safety of artificial sweeteners. Pandora's box was opened.

In June 1965, cyclamate claims as weight-reducing aids were shattered. Instead of keeping weight down, artificial sweeteners appeared to increase the craving for real sweets. For diabetics who departed from their customarily prescribed low-carbohydrate diet and became dependent on artificially sweetened foods, grave health consequences could result.[13] Researchers at the Frances Stern Food Clinic in Boston had conducted studies on a hundred diabetic women to learn whether the artificial sweeteners helped them to stay on a carbohydrate-restricted diet. The conclusion was that "there is little basis for implying that adherence or non-adherence to a carbohydrate-restricted diet is related to the use of a non-caloric sweetener." [14]

Medical journals were publishing clinical reports about cyclamates' effects on humans, with a wide spate of symptoms, ranging from dizziness, rash, itch, ear fluttering, and photosensitivity, to sleep interference. Some reports suggested possible links between cyclamates and retardation in children, and mental aberrations.[16] Cyclamates interfered with many types of drugs, reducing effectiveness of an oral anti-diabetic; increasing hypoglycemic effects of chlorpropamide; potentiating diuretic effects of thiazide with resulting excessive potassium loss; potentiating anti-coagulant effect of coumarin; binding strongly to plasma protein resulting in displacement of other drugs similarly bound; and reducing absorption of lincomycin, an antibiotic used to fight bacterial infection. Ironically, cyclamates were used to flavor some antibiotics.[16]

By September 1965, hard evidence was surfacing regarding cyclamates' health hazards. Results were announced from a nine-month study in which rats were fed cyclamate at various dietary levels. Cyclamate decreased conception ability, produced stunted offspring, and caused litter deaths. Rats on a diet containing 5 percent cyclamate for nine months suffered 12 percent growth impairment; at 10 percent cyclamate, 50 percent growth impairment. The female rats appeared to bear normal offspring, but a few weeks after the young were weaned their growth rates dropped. Among the offspring of rats fed the 5 percent cyclamate diet, 15 percent became stunted; 35 percent of the offspring, of females on 10 percent.[17]

This study had been conducted by the Wisconsin Alumni Research Foundation (WARF), funded by The Sugar Research Foundation, Inc. The sugar interests had voiced concern over the surging popularity of artificial sweeteners, which obviously were cutting deeply into sugar sales. While data from these studies were not challenged, motives were.

The sugar foundation continued its financial support of WARF's studies, and by 1967, very damaging evidence was revealed in rat studies. Cyclamates were found to cause, among other effects, increased adrenal weight and structural changes in the adrenals, testes, kidneys, and pancreas; pituitary gland damage; a high incidence of kidney stones; fewer white cells in the blood of male rats fed 10 percent cyclamate; and retarded growth rate. Adult rats on restricted feed and 5 percent cyclamate produced a first litter that died within 5 days; the second litter, within 7 days. Conception was suppressed totally in adult rats fed 10 percent cyclamate.

Charges and countercharges followed. Abbott scientists reported that no other researchers were able to duplicate WARF's results, and that definitive independent research had not been conducted to date.[18]

WARF then released Japanese studies, reported earlier in the year, which showed teratogenic effects (inducing birth defects) in rats fed cyclamate during the first two weeks of pregnancy. Abbott scientists and FDA pharmacologists responded by saying that these Japanese studies had not been duplicated either.

Pressured by the sugar interests and by congressmen who had been prodded by this group, FDA issued a contract to help support animal and human cyclamate studies at Albany Medical College, led by Dr. Frederick Coulston. The studies, to receive additional financial support from two cyclamate producers, Abbott Laboratories and E. R. Squibb, and from The Sugar Research Foundation, could hardly be regarded as independent research, but would enlarge understandings. Cyclamates were to be tested with rats and monkeys, as well as with prison volunteers, for possible effects on enzyme systems, liver tissue, and reproductive functions including sperm production.

While studies were progressing, Coulston admitted that, as yet, cyclamates could be neither exonerated nor condemned, but "use should be restricted to persons who really need to cut out sugar." Otherwise, he said, "it is possible their use will get out of hand." [19] The greatest concern was the welfare of children, who might drink four or five bottles of diet cola daily, along with artificially sweetened candies and cookies. Up to that period, only one study with children had been reported, and results were inconclusive.

Another aspect was given attention. Cyclohexylamine (CHA) is the basic chemical from which cyclamates were manufactured. CHA, used in many manufacturing processes and in insecticides, was known to cause dermatitis if applied to the skin, or even to lead to convulsions if inhaled. This toxic compound could be found, too, as an impurity in cyclamates or produced during food processing in cyclamate-containing products. But nobody suspected that CHA might be produced within the body after cyclamate consumption.[20]

Originally, Abbott scientists claimed that cyclamates are not metabolized by the body, but simply are excreted as cyclamates. However, Japanese researchers reported that some individuals excrete CHA, indicating that the body can break down cyclamate into at least one metabolite. Later work at Abbott Laboratories, as well as at Albany Medical College, supported the Japanese finding. CHA was found in the urine of about 12 percent of all persons consuming cyclamates, even at low consumption levels. In animal experiments, the metabolite was found in cyclamate-fed dogs and in a strain of rats.[3] (Later, it was learned that CHA, itself, breaks down to an even more toxic form, dicyclohexylamine.)

Additional disturbing news, reported by Dr. Marvin Legator, chief, Cell Biology Branch, FDA, was that in both test tube and animal studies, even in moderate amounts, CHA caused significant chromosome breakage. Hence, the metabolite was a suspected mutagen. Also, it was a suspected carcinogen.[21] The incidence of chromosomal breakage in both sperm and bone-marrow cells was directly related to CHA levels. The breaks, occurring mostly in one portion of the chromosome, were likely to have carcinogenic effects. Chromosome breaks occurred at relatively low CHA levels, comparable to levels of CHA consumed by humans.[22]

Other studies showed that CHA and cyclamates were transported from the pregnant rhesus monkey through the placenta into the developing fetus. While cyclamates were transported only to a limited extent, CHA crossed the placenta freely.[23]

In Albany Medical College tests, 9 of 24 prison volunteers consuming cyclamate were plagued by severe, persistent diarrhea; the higher the dose, the worse the condition. Of the 24 men, 17 converted cyclamate to CHA; and in some, the metabolite was found to cause vascular constriction and increased blood pressure. This feature was regarded as a potential risk for cardiac patients with high blood pressure who might be using cyclamates for weight control. Also, marked increase in protein-bound iodine was noted. Patients could be misdiagnosed and treated for hyperthyroidism, while in reality they might be showing signs of high cyclamate consumption.[24]

Scientists voiced nagging questions: Why were these findings not found when cyclamates were first tested, before being given approval? Were initial safety tests incomplete, slipshod, or both? Although the drug companies and scientific investigators had complied with FDA standards, the newer findings illustrated gaps and inadequacies of testing protocol.

In 1967, FDA requested a third cyclamate safety review by the Food Protection Committee, since newer toxicologic information and increased cyclamate usage made it necessary to re-evaluate. In 1957, the annual sale of cyclamates was 2 million pounds; by 1967, the annual sale reached 15 million pounds, and this volume was anticipated to reach 21 million pounds by 1970. Three fourths of the entire population was consuming non-nutritive sweeteners. Clearly, FDA had tolerated uncontrolled usage, and the situation was out of hand.[25]

The following year, the Food Protection Committee released its new review and repeated that "totally unrestricted use of cyclamates is not warranted at this time." The committee noted the lack of specific information about consumption patterns: the range of intake by different age groups; the extreme upper levels of these ranges; and the number of individuals consuming such levels. The committee found it difficult to evaluate some of the ongoing, incomplete toxicologic

studies and recommended additional research to clarify existing animal studies. The committee reiterated its past position,
and added that intake of five grams (5000 milligrams) or
less per day of cyclamates by adults should present no hazard
but the daily intake should be based on body weight. The
committee suggested that an adult should consume no more
than 70 milligrams (mg) of cyclamates per kilogram (k) of
body weight daily. This recommended limit of 70 mg/k was
less stringent than that of the Food and Agriculture Organization and World Health Organization (FAO-WHO) of the
United Nations, which had suggested an upper limit of
50 mg/k.[26]

FDA scrutinized the new report carefully since it contained
implications for regulatory policy. Also, tardily, the agency
had begun systematically to build some expertise with artificial
sweeteners. Dr. John J. Schrogie of FDA's Bureau of Medicine
and Dr. Herman F. Kraybill of FDA's Bureau of Science
were asked to evaluate the latest safety review. These men
judged the review superficial and said it represented

> a largely uncritical review of the available material. Conclu-
> sions of the studies are included without proper regard for the
> quality of methodology originally used. Studies lacking ade-
> quate statistical design are given equal weight with sounder
> studies; many clinical and epidemiological studies yielding
> questionable conclusions are given uncritical acceptance. Be-
> cause such conclusions are included, possibly erroneous results
> gain added stature and interpretive errors are perpetuated.

Schrogie and Kraybill suggested that the committee had placed
undue reliance on usage figures supplied by an industry survey
that was "fraught with great hazard of error" and chose to
ignore some pertinent teratogenic and mutagenic studies.[9]

According to Philip Boffey, who made a detailed study of
NAS's cyclamates reviews, the only geneticist serving on the
Food Protection Committee's 1969 review was Dr. James F.

Crow, who had concluded that the evidence added up to a "fairly strong case that cyclamates represent a mutagenic hazard to man." Boffey reported that Crow asked that his name not be included in the report because he could not endorse it. No hint of Crow's input appeared in the report. His views were not afforded space as a minority report and the Committee never acknowledged his dissension. Crow was listed merely as "consultant" on the report, a designation that conveniently allowed the Committee to imply that he had contributed to the report without, in fact, requiring his signature.[9]

As a result of the new review, FDA lowered its previous recommended upper cyclamate limit to 5 grams daily for an adult. Then, in April 1969, the agency urged people using cyclamates to reduce this limit to 3.5 grams (gm) daily for adults and 1.2 gm daily for children. These limits were applicable to a 154-pound adult or a 54-pound child. Since some diet soft drinks contained up to a gram of cyclamate in one 12-ounce bottle, young children easily could exceed the recommended maximum. These suggestions and revisions simply confused the public. Furthermore, consumers found it impractical, if not impossible, to control cyclamate use on the basis of label information on artificially sweetened products. FDA proposed a new regulation to require that such foods and beverages include a declaration of cyclamate content in milligrams, along with a statement specifying upper limits recommended for adults and children. This labeling recommendation was never put into effect.[26]

Labeling was a minor issue in the cyclamate affair as controversy continued to rage. The sugar interests kept sounding alarms about cyclamate dangers. Abbott Laboratories appealed to processors of artificially sweetened foods and beverages to assist in a public relations campaign defending cyclamates' safety.[27]

Meanwhile, Dr. George T. Bryan, a tumor expert and

cancer researcher, reported that, to date, no attention had been given to the urinary bladder as an organ susceptible to cyclamates' cancer-inducing potential. Bryan had devised a technique that demonstrated this potential. The technique was highly sensitive, reproducible, and predictive of cyclamates' cancer-inducing property in the bladder. Bryan surgically implanted pellets, composed of either sodium cyclamate and cholesterol, or pure cholesterol as control, into mouse bladders. The bladder exposure to cyclamate was very brief, since half of the compound disappeared from the pellets within about an hour. The animals in each group were allowed to live for 13 months, then were sacrificed and examined. Bladders of any animals who had survived more than 175 days were examined microscopically. The cyclamate-treated mice had a high incidence of bladder cancer, while the control group did not. Bryan claimed that the bladder-cancer induction demonstrated by pellet implants predicted the cancer-inducing property of orally ingested cyclamates.

Although many scientists supported Bryan by endorsing the validity of the pellet implant technique, FDA dismissed it as controversial and inappropriate.[28]

Within FDA, disquieting data mounted. In addition to Legator's findings, Dr. Jacqueline Verrett, another FDA scientist, had worked with 10,000 eggs containing chicken embryos injected with cyclamate 96 hours after incubation, a period corresponding to the critical early months in human pregnancy. When hatched, many of the chicks were deformed, having twisted spines, underdeveloped eyes, dwarfed and missing wings and legs, and most frightening, flipper arms and legs, which were the familiar characteristics of thalidomide babies.

Early in October 1969, in a dramatic presentation, Verrett displayed some of the deformed chicks before national TV cameras. She implied that pregnant women should avoid cyclamate consumption.

Verrett's presentation was followed swiftly by FDA Commissioner Dr. Herbert L. Ley, Jr., requesting NAS to evaluate Legator's new data and Verrett's findings. One HEW researcher expressed fear that "panic would force the federal government to make decisions based on practical necessity rather than scientific evidence. Not only the industry but millions of Americans could suffer." Events moved rapidly.[27]

Ley's request for a new NAS review prompted HEW Secretary Robert Finch to assail FDA's varying cyclamate assessments and accuse the agency of "waffling." While FDA was attempting to formulate official actions, Abbott sent new data to the agency showing that rats fed high cyclamate levels during most of their lifespan developed cancer. This information was reviewed promptly by the National Cancer Institute (NCI) and by the NAS's Ad Hoc Subcommittee on Cyclamate Safety.[29]

The Delaney Clause prohibits the use of known cancer-inducing agents in the food supply. On October 18, 1969, in a move that surprised the general public but not the cyclamate industry, Finch invoked the Delaney Clause and ordered cyclamates removed from GRAS and banned, with an orderly market phase-out by February 1, 1970. The U.S. ban was followed by similar bans of cyclamate-containing foods and beverages in Great Britain, Canada, Sweden, and elsewhere.[30]

Due to FDA's spineless policy and inept fumblings with the cyclamate issue, Ley was forced to resign. In a press interview he candidly admitted, "The thing that bugs me is that the people think the FDA is protecting them. It isn't. What the FDA is doing and what the public thinks it's doing are as different as night and day."[31]

In retrospect, cyclamates should never have been granted GRAS status, which automatically bestowed the privilege of exemption from classification and regulation as a food additive. If cyclamates had been classified as food additives, safety testing would have been mandatory. At an early stage, when questions were raised about cyclamate safety, FDA had an obligation to remove cyclamates promptly from GRAS. If the agency had acted properly, the ban would have been instituted decades earlier, since the necessary proof of safety was lacking. Countless numbers of people would have been spared from needless carcinogenic exposures.

The cyclamate ban announcement resulted in wild stock market trading. Frantic speculators attempted to unload diet product stocks before prices dropped precipitously, and to buy sugar stocks before their prices soared. Heavy cyclamate consumers went on shopping sprees and hoarded cyclamate-sweetened foods and beverages. Seven panicky consumers filed suit in a federal court to halt the ban.[32]

Federal actions were termed "arbitrary, unreasonable, and capricious," by *Barron's,* which typified most reactions from the business and financial communities. A few scientists branded HEW's curb as a sweeping decree after a "hurried meeting" and "on the basis of experiments employing only 12 rats." This charge was untrue.[33]

Earlier studies had suggested a link between cyclamates and cancer. At least 21 published cyclamate studies, mainly from 1951 to 1968, and reviewed by NAS's committee, turned up possible cancerous effects. However, the committee judged that the tumors formed not much more frequently in cyclamate-fed than in control rats, and on that basis dismissed the findings as insignificant.

Among the 21 studies was a crucial one, conducted by FDA from 1948 to 1949. A large number of rats were fed a cyclamate-saccharin mixture over a two-year life span. When the 1969 bladder cancer report was received from Bryan, FDA assigned two pathologists from its Bureau of Science to re-examine the earlier data and microscopic tissue slides. They concluded that, indeed, there *had* been reason to suspect cyclamates of cancerous effects as early as the late 1940s. The NAS's committee had never been given specific data on these early studies regarding incidence and variety of possible tumors in the rats. Hence, the committee's suspicions had not been aroused.[34]

Another study, from 1951, demonstrated that cyclamates could induce ovarian tumors. While the study was available to NAS's committee, it was not released publicly until after Finch's October 18, 1969, cyclamate ban order.[35]

A month after Finch's actions, the economic blow was eased for industry. By legal legerdemain, FDA would no longer classify cyclamates as food additives but would reclassify them as drugs. This deft maneuver skirted the Delaney Clause, which is applicable to foods but not drugs. Cyclamates now would be considered as over-the-counter non-prescription drugs and labeled as drugs.[36] Cyclamates could be used in foods and as sugar substitutes in liquid and tablet forms, with labels stating the cyclamate content of average servings. The ban against cyclamate use in diet soft drinks and other beverages would remain in effect.

The reclassification attempt backfired. FDA was willing to approve cyclamates and cyclamate-containing foods if companies filed "abbreviated" new drug applications, waiving safety and efficacy proof customarily required. Representative L. H. Fountain (D.–N.C.), who chaired congressional hearings on cyclamates, declared such approval as "ill advised" and illegal. Fountain demanded that such approval be rescinded promptly.[37]

Safety had not been proven. On the contrary, by then, there was a sizable body of evidence demonstrating harmfulness. Efficacy had not been proven. Evidence showed that cyclamates lacked value in weight-control programs. Dieters depending on artificial sweeteners tended to make up the calories by consuming extra portions of foods and beverages and lost weight no faster than those who used sugar moderately.[34]

Early in 1969, rat experiments in Holland had demonstrated that cyclamates *stimulated* appetite and led to weight *increase*.

The researchers had concluded, "If this is true for man, then weight reduction will be more difficult if the diet contains cyclamates."[38]

With neither safety nor efficacy proven, FDA was forced to reimpose a total cyclamate ban, issued on August 14, 1970, and effective September 1, 1970. A nationwide survey conducted during the month after the cut-off date showed that half the stores were selling foods and beverages containing the illegal cyclamates. FDA took no vigorous action to enforce the order and merely relied on voluntary compliance.[39]

On October 8, 1970, the House Committee on Government Operations justifiably and sharply criticized both FDA and HEW for improper cyclamate regulation. The committee charged that by 1966 a genuine difference of opinion existed among qualified experts regarding cyclamates' safety, and by then, FDA had been obliged to delete cyclamates from GRAS, and ban them.

The committee's findings were that

> FDA failed for several years to protect the public against possible health hazards associated with cyclamates despite a clear legal obligation to do so. FDA aggravated the consequences of its inaction by permitting the use of cyclamates in food to reach massive proportions. FDA attempted to permit the continued marketing of cyclamate-containing products through illegal regulations and procedures. The decision to permit the continued marketing of cyclamate-containing products was made by the Secretary of HEW, not by FDA. HEW used an outside advisory body to make recommendations on matters that had already been decided, involving a basic issue which the advisory body was not qualified to decide. NAS-NRC panels that considered the safety of cyclamates in food were not asked to provide the basic information necessary for determining if cyclamates should remain on the GRAS list.

After firing these volleys, the committee recommended that, in future, FDA and HEW should "take prompt and effective action to guard against repetition of the mistakes made in the regulation of cyclamates." Perhaps the committee already had a premonition that the two agencies would repeat the same folly in their future handling of the saccharin issue.

The committee further recommended an immediate and objective review of any GRAS list substance whenever its safety was questioned and its prompt removal from the list. The agencies were advised to confine their use of scientific advisory bodies to consideration of clearly defined issues that are within the competence of such bodies. Also, such groups should not be asked to advise on matters that the regulatory agency, FDA, has the capacity to resolve. Food and drug safety questions should be decided "strictly in compliance with legal and scientific requirements and without regard to economic and other extraneous considerations."[40]

But economic considerations continued to play a role. Shortly after the total ban, producers of cyclamate-containing foods and beverages attempted to recoup their losses through different strategies. They dumped products for whatever price they could get. Or, they donated them to charity and were rewarded with substantial tax savings. Throughout America, high mounds of diet soda were stacked in thrift shops, frequented mainly by the poor; and in third-world countries, the poor also became recipients.[41] One shipment of 60,000 cases of a low-calorie drink, intended for calorie-starved Laotian refugees, was stopped by the actions of a U.S. congressman who considered the gesture "cheap and cruel." Taxpayers would have picked up the $42,000 shipping tab and the company would have enjoyed a tax break.[42]

By 1972, in response to activity described as "an enormous amount of lobbying," a bill was introduced in the House of Representatives to reimburse manufacturers and distributors who claimed economic losses caused by the cyclamate ban. The bill opened the way for claims estimated between $120 and $500 million against the Federal government. The unprecedented bill proposed that the government should pay damages, even though no one charged the government with any wrongdoing. The cyclamate users' premise was that they had relied in good faith on cyclamates' safety because of its inclusion on GRAS. Although the bill passed in the House, it was killed in committee hearings in the Senate.[43] However, as recently as the summer of 1980, the California Canners and Packers filed suit seeking damages due to financial losses suffered from the 1969 cyclamate ban.[44]

After the cyclamate ban, studies were continued in various research laboratories, here and abroad. In newer tests, researchers used more animals, different species, higher cyclamate doses, and used longer test periods than in earlier studies. The newer tests failed to confirm earlier research implicating cyclamates as carcinogens, although other adverse effects were discovered.

By 1973, the Calorie Control Council, a trade association comprised of producers of artificial sweeteners and processors who used them, spearheaded a drive to lift the ban and put cyclamates back into production and marketing. The drug industry followed these activities with interest.[45]

In November 1973, Abbott Laboratories petitioned FDA to reinstate cyclamates as safe additives.[46] By mid-February 1974, Abbott had complied with FDA's request for data by dramatically placing in public view at the Hearing Clerk's office some 400 studies, bound in 17 weighty volumes, along with supplementary binders. FDA assigned a team of toxicologists, nutritionists, and chemists to review the voluminous data before passing them along to NAS's Food Protection Committee. Both Abbott and FDA were smarting from earlier criticisms and, this time, were attempting to have a thoroughgoing analysis and review.[47]

By September 1974, FDA rejected Abbott's petition on grounds that the data were "inconclusive." FDA reported that, as yet, questions about cyclamates' cancer-causing potential were unresolved. FDA requested further scientific studies, among which were duplication of the original rat studies that led to the cyclamate ban and additional data on possible effects on the reproductive organs and the cardiovascular system.[48]

Abbott maintained that further scientific studies were "neither reasonable nor required by scientific judgment." Abbott hoped to resolve all FDA objections at an unprecedented public conference scheduled in November 1974 between company and government scientists. At the meeting, FDA took no action but informed Abbott representatives that the company's latest presentation would be considered and discussions would continue by mail.[49]

By March 1975, FDA reversed its position and told Abbott that it need not conduct additional time-consuming safety tests. The agency would make a decision based on the available evidence. At the same time, FDA requested NCI to convene a blue-ribbon panel to conduct a fair evaluation of all current evidence bearing on the issue of cyclamates' cancer-causing potential, and to make recommendations. Abbott interpreted this news as encouraging.[50]

Further heartening news followed ten days later, at a NAS forum, sponsored by U.S. Public Health Service (PHS), FDA, and HEW. Dr. Philip Handler, NAS president, charged that the experiments leading to the cyclamate ban were badly designed, inconclusive, and should not have warranted any action at that time. Handler's remarks echoed those of Sveda, the discoverer of cyclamates, who also participated at the forum, and who, ever since the ban, had been protesting vigorously and working energetically to overturn it.[51]

Scientists selected to participate in the forum were predominantly proponents of cyclamate reinstatement, and the well-publicized meeting was hardly representative of a balanced view. Handler, himself, had served from 1964 to 1969 as a member of the board of directors of a company that used cyclamates in its brand of sweetening agents. Handler had told Philip Boffey, a critic of NAS's industry affiliations, that he had resigned his directorship and sold his company stock when he assumed presidency of NAS in 1969. Handler also said that the company had been marketing the sweetening agent before he joined the board and his role was to maintain liaison with the research departments that were not involved with cyclamates. Nevertheless, Handler was associated with a company that participated in the phenomenal increased cyclamate use. Boffey concluded that Handler presumably "retains the conditioning of a corporate director who would tend to regard the products of his industry as beneficial and efforts to ban them as unreasonable." [9]

NCI's panel could not reach a unanimous agreement in its review. By the middle of December 1975, in a preliminary conclusion, three of the five members tentatively agreed that cyclamate is not a strong carcinogen. The remaining two hedged, stating that cyclamate may be a weak carcinogen but there is no way to prove absolute safety.

By January 1976, the panel report went to NCI's director who, after review, sent it to FDA by mid-March. After a nine-month review, the final report concluded that present evidence, from animal studies, did not establish the carcinogenicity of cyclamates or CHA. Due to lack of epidemiological data, no conclusion could be made about cyclamates' cancer-causing potential in humans. But the panel was concerned about the increased urinary tract tumor incidence noted in several studies with cyclamate-fed animals. Thus, FDA continued to be confronted with the dilemma that the agency had hoped would be resolved by NCI's panel.[52]

While that panel was deliberating, FDA scientists reviewed other cyclamate studies. The agency concluded that large amounts of cyclamate in test animals can affect growth and reproduction, cause testicular atrophy, and elevate blood pressure. Even if there were no cancer issue, based on these findings, any permissible safe level would be too low for practical cyclamate use as a sweetener.[53]

On May 11, 1976, FDA advised Abbott to withdraw the petition to remarket cyclamates. As yet, the cancer question was unresolved and other issues needed clarification. Several newer studies suggested that cyclamates, at doses approximating ordinary uses, caused human genetic damage.[54]

Abbott informed FDA that it would not withdraw the petition. This action forced FDA to issue a formal denial, after which Abbott could request a public hearing. For the next two years protracted hearings were held. Repeatedly Abbott petitioned, and repeatedly, the petitions were denied.[55]

By 1978, an FDA administrative law judge, Daniel Davidson, reviewed the case and concluded that cyclamates had not been proved safe and therefore the ban could not be lifted. Abbott continued to re-petition. On June 26, 1979, hearings were reopened for additional evaluation before a final decision to approve or disapprove cyclamates' use.[56] By early September 1979, no decision had been made. To force action, Abbott filed legal suit against FDA. In response, the agency promised a final decision within a year, suggesting September 15, 1980, as a target date. On February 4, 1980, Judge Davidson again rejected Abbott's petition, stating that the data did not prove safety. Again, Abbott re-petitioned. On September 4, 1980, FDA denied the petition. By then, evidence including mouse and rat studies showed increased numbers of tumors in lung, liver, bladder, and lymph systems in cyclamate-fed animals.[57] Other animal studies showed that cyclamates could adversely affect chromosomes, suggesting that the sweetener could cause inheritable genetic damage that could lead to Down's syndrome, mental retardation, and altered metabolism.[58]

After this petition denial, Abbott finally decided not to make further appeal. The battle was over. The cyclamate issue was dead, but the saccharin issue was heating up.[59]

Saccharin

The story of saccharin is inextricably interwoven with that of cyclamates. Some developments with both artificial sweeteners are so strikingly parallel as to give one a sense of déjà vu.

Saccharin, like cyclamates, was discovered accidentally, and by current standards, also due to sloppy procedures in a chemical laboratory. In 1879, Dr. Constantin Fahlberg, a graduate of the University of Leipzig, worked in a Johns Hopkins laboratory under Dr. Ira Remsen. During work on the oxidation of o-toluenesulfonamide, Fahlberg synthesized o-benzosulfimide. One day, working with this compound in the laboratory, he munched on bread that tasted incredibly sweet. Suspecting that he had contaminated the bread with the compound, he analyzed the compound and, finding that it was 300 to 500 times sweeter than sucrose, he named it saccharin.

Shortly after, Fahlberg returned to Germany and obtained patents for saccharin and its commercial production methods. In the early 1880s, Fahlberg and his coworkers began commercial production in Germany of sodium and calcium salts of saccharin. In the United States the Monsanto Chemical Company, founded with the express purpose of manufacturing saccharin, began commercial production in 1902. Soon other American companies entered the market.[60]

From saccharin's inception, it was regarded as a drug, possibly of limited usefulness for certain dietary purposes, such as diabetes. As early as 1890, a decade after saccharin was synthesized, the French Commission of the Health Association decreed saccharin harmful and forbade its manufacture or import. This was followed by the German government's limiting its use as a drug and expressly forbidding its addition to any foods or beverages. Similar regulations were enacted in Spain, Portugal, Hungary, and elsewhere.[61]

In the United States, however, attempts to keep saccharin

out of the general food supply were unsuccessful. American food and beverage processors favored the use of saccharin, since the sweetener was cheaper, sweeter, and easier to handle than table sugar. By 1907, the American marketplace was flooded with saccharin-sweetened canned fruits and vegetables, candies, soft drinks, and bakery products.

Harvey W. Wiley, M.D., who had conducted the unsuccessful campaign against glucose discussed in Chapter 2, attempted to keep saccharin out of the food supply. Wiley warned that "saccharin is a noxious drug and even in comparatively small doses it is harmful to the human system." Wiley regarded the use of saccharin as a sugar replacer not only as harmful but also as a practice of adulteration and deception.

Wiley attempted, unsuccessfully, to enforce a saccharin ban. His Federal agency brought action against Monsanto and presented what Wiley considered satisfactory evidence regarding saccharin's harmfulness. But twice the case resulted in hung juries. The ban was appealed, and hearings were held in the presence of President Theodore Roosevelt. One corn processor boasted that his company had saved $4000 one year by using saccharin as a sugar replacer. Wiley retorted, "Everyone who ate that corn was deceived. He thought he was eating sugar, when in point of fact he was eating a coal tar product totally devoid of the food value and extremely injurious to health."

What followed, in a celebrated exchange between Wiley and Roosevelt, was described by Wiley in his autobiography:

> Turning to me in sudden anger the President changed from Dr. Jekyll to Mr. Hyde, and said, "You tell me that saccharin is injurious to health?" I said, "Yes, Mr. President, I do tell you that." He replied, "Dr. Rixey [Roosevelt's personal physician] gives it to me every day." I answered, "Mr. President, he probably thinks you may be threatened with diabetes." To this he retorted, "Anybody who says saccharin is injurious to to health is an idiot."

Roosevelt's remark broke up the meeting. According to Wiley, the incident was "the basis for the complete paralysis" of consumer-protective food regulations.

Roosevelt ordered the establishment of a referee board of consulting scientific experts to review saccharin's safety. A fair evaluation might have been possible with unbiased scientists. But none other than Dr. Ira Remsen, co-discoverer of saccharin, was chosen to head the board and to select other members. Despite the glaring conflict of interest, Remsen did not disqualify himself. He served and packed the board with cronies. Wiley viewed the board's composition as "the worst crowd of adulterators that ever infested a republic."

The Remson board not only exonerated saccharin but usurped the functions and powers of Wiley's department. The board continued to hand down decisions that unraveled Wiley's accomplishments for protecting consumers against hazardous substances. Instead of regulating control of an increasing number of chemical additives that were being introduced into the American food supply, the board removed most constraints.[62]

Saccharin was in, and before long, Wiley was out. The Remsen board's decisions cast a long shadow and ultimately bore some responsibility for many crucial issues that inevitably developed into major crises several decades later.

Judged by present standards, early saccharin safety tests were crude, unsystematic, and lacking controls. Nevertheless, many reports supported Wiley's viewpoint.[63]

The first suspicion raised about saccharin's possible carcinogenicity was in 1948, in a chronic toxicity test by FDA. Three FDA scientists reported in 1951 that saccharin at certain levels induced a high incidence of unusual combinations of cancers. Inexplicably, FDA chose to ignore their own data and retained saccharin as GRAS.[64]

In 1955, with NAS's first review of artificial sweeteners, interest was mainly focused on cyclamates, which were relatively new substances. Saccharin was scrutinized less since its long-time use without apparent ill effects appeared to be no cause for concern to the panel. The existing earlier studies and medical reports were either discredited or ignored. A similar attitude toward saccharin prevailed in NAS's later reviews in 1962 and 1968.

Using a bladder pellet implant technique in a single experiment with 12 mice, in 1957, researchers found that "pellets containing saccharin induced a significant incidence of bladder tumors." Despite this report, saccharin remained GRAS.[65]

In October 1969, FDA announced that saccharin would be "restudied," noting that although its use spanned more than half a century "very little valid information" was available concerning its safety. An agency spokesman reported that some further testing would be needed to insure that saccharin merits its "present clean bill of health." FDA proposed that agency scientists review all the scientific literature to learn what had been done and what gaps in knowledge might exist.[66] The Monsanto Company provided FDA with all its data, and began two-year animal feeding studies with saccharin.[67]

When FDA had banned cyclamates on the basis of Abbott's studies, there had been some justified criticism. The studies were flawed since the animals had been fed a mixture of cyclamates and saccharin. It was impossible to determine whether the cancers had been induced by the cyclamates or saccharin or the combination of the two, yet cyclamates were banned and saccharin was continued.

Meanwhile, Dr. George T. Bryan, who had conducted the cyclamate-pellet implants, had applied the technique in new studies with saccharin. By the end of November 1969, Bryan notified FDA that saccharin-treated mice developed a high incidence of bladder cancer. By mid-January 1970, an FDA team visited Bryan's laboratory, reviewed all slides and data, and confirmed Bryan's conclusions. Viewed microscopically, the saccharin-induced bladder cancers were qualitatively more severe than the cyclamate-induced tumors. They were larger, appeared deeper in the bladder tissues, and apparently were in a more advanced stage. The saccharin-induced tumors were as malignant in appearance as those induced in mice, rats, and dogs by feeding them well-recognized powerful urinary bladder carcinogens.

Bryan acknowledged that a direct saccharin cancer hazard for humans was not yet established, but strong suspicions had been raised. Bryan cautioned, "It may take many years before it is known exactly how dangerous the substance is, and until then its use should be restricted to those who need it for medical reasons." [68]

Primarily due to Bryan's findings, FDA announced in March 1970 that NAS was requested to review saccharin. After deliberation, by late July 1970, NAS's committee released its conclusions. Although newspaper headlines proclaimed "NO HEALTH HAZARD FOUND IN SACCHARIN" and "PANEL FINDS SACCHARIN SAFE," further reading of the articles revealed the opposite. Far from exonerating saccharin, the NAS's committee acknowledged that available information was incomplete and urged further laboratory experiments. Questions needed to be answered, especially since cyclamates were unavailable, and saccharin use was likely to increase. The committee also wanted information about saccharin's effects with other drugs. Animal studies had shown that saccharin interacted with certain medicines, including insulin, so that diabetics might be at particular risk. Saccharin crossed the placenta in the rhesus monkey during late pregnancy and became distributed in many fetal tissues. In contrast to its rapid excretion from the maternal kidney, it cleared very slowly from the fetus. The committee concluded, nevertheless, that "on the basis of available information, the present and projected use of saccharin in the U.S. does not pose a hazard." [69]

Despite this assessment, FDA still faced unresolved safety questions. By mid-September 1970, FDA announced a possible restriction on saccharin use. The agency hoped to hold saccharin consumption to then-current, or perhaps lower, levels as a precautionary measure should human health hazards be demonstrated later.

In retrospect, FDA officials recognized that many difficulties in the cyclamates affair could have been avoided if, early on, the agency had imposed restrictions on cyclamates' use. The agency wanted to avoid repetition of this mistake, since saccharin was being added to foods and beverages in increased and unregulated quantities, just as cyclamates had. After the cyclamates' ban, it had been thought that saccharin's distinctly bitter aftertaste automatically would limit its use, but processors found other compounds to mask the bitterness, and more saccharin was being added to products.

Nine months elapsed before FDA finally suggested on June 25, 1971, some curbing actions. The agency proposed to drop saccharin from GRAS and to place specific limits on its use. The agency formulated an unprecedented plan to place saccharin in a provisional food additive category to allow "interim" use pending the outcome of more research, and "freezing" saccharin use to existing levels. Since completion of some ongoing research was expected to take as long as ten years, the interim status could serve as a convenient delaying tactic. No safety assurance could be given for the existing high use levels, but the freeze would prevent economic disruption.

Also, despite the fact that no "safe" level had been determined, FDA recommended that saccharin use be limited to about one gram a day for an adult and proportionately less for a child. The limit, coincidental or not, matched the average one-gram daily intake of a 150-pound heavy saccharin–consuming adult. Most 16-ounce bottles of diet soft drinks contained nearly 0.1 grams of saccharin, and some as much as 0.2 grams. Heavy users of saccharin-containing foods and beverages, who also used tabletop saccharin packets to sweeten foods, easily could exceed the recommended limit.[70]

Furthermore, the recommended limit was based on a mathematical error in NAS's calculations, with a misplaced decimal point. The committee intended to use a time-honored hundredfold safety margin: one-hundredth of the dose level at which no apparent harm is observed in animals fed the substance. The error resulted in a thirtyfold safety margin, considered by toxicologists as uncommonly low, but by FDA as reasonably safe.[71]

Seven months elapsed before FDA removed saccharin from GRAS, effective February 1, 1972, and issued an interim provisional regulation restricting saccharin use pending completion of safety reviews. FDA's actions were part of a larger program, long overdue and mandated by Congress, to review more than 600 GRAS list substances.

FDA's actions with saccharin were precipitated by the nearly completed two-year rat feeding studies, conducted by the Wisconsin Alumni Research Foundation and funded by the International Sugar Research Foundation, Inc. Preliminary reports indicated that some male rats fed saccharin at a 5 percent level in the diet developed bladder tumors that appeared to be malignant. If malignancy was confirmed by government scientists and NAS's committee, FDA would be obliged to invoke the Delaney Clause and ban saccharin. In addition to the WARF studies, 11 more studies were nearing completion in the United States at FDA and NCI, as well as in West Germany, Holland, and Canada, conducted with rats, mice, hamsters, and rhesus monkeys. Also, epidemiologic studies had been launched in England, Germany, and the United States.[72]

If saccharin were banned, the economic stakes were high. In the early 1950s, only about twenty thousand pounds of saccharin were used yearly in the United States. By 1976, the figure had risen to about seven million pounds for food use, produced here and supplemented by substantial imports from Japan and Korea. Only 15 percent of that total was used in special purpose dietetic foods. (Contrary to popular belief, only about a third of all diabetics use saccharin regularly.) The general public was consuming 75 percent in diet soft drinks and 10 percent as table top sweeteners. In restaurants, saccharin-containing packets became commonplace, accompanying sugar packets in the sugar bowl. Trade journals reported that food service operators could cut sweetening costs by offering saccharin. Restaurateurs were urged to use saccharin. "Promote iced tea sweetened with [brand name] as a perfect diet drink, all year round. Patrons love fewer calories while you save from high cost of sugar." Magazines read by homemakers carried advertisements featuring recipes using saccharin. "Sinlessly sweet snackin' squares," sweetened with saccharin, were described as "unique, low-calorie snacks in fabulous flavors. So delicious they taste like no-no's." [73] Non-food uses of saccharin as a flavoring agent appear in tobacco and cigarette paper; toiletry articles such as dentifrice, mouthwash, lipstick, and cosmetics; and medical preparations including cough syrup, children's aspirin, and antibiotics for pediatric use.[74] Saccharin is even used in animal feed. At no point has FDA ever engaged in a public education campaign, as HEW has done with tobacco, to lower or discourage general saccharin use.

On February 27, 1973, FDA announced that in its own carefully controlled studies, suspicious bladder tumors were found in rats fed saccharin at 7.5 percent of the diet. It was thought that the tumors might have been caused by impurities in the saccharin, or by mechanical bladder irritation caused by the high saccharin level. FDA announced that it would take no action, pending NAS's planned review of the study.[75]

FDA researchers were examining saccharin not only for any cancer-inducing potential but also for possible reproductive hazards, such as birth defects and genetic changes. On May 21, 1973, FDA reported that the pathology review of its study showed "presumptive evidence" that saccharin caused cancerous bladder in rats. In addition, saccharin retarded the growth rate.[76]

FDA was required to act by June 30, 1973, to end, extend, or modify saccharin's interim status. Clearly, NAS would be unable to meet this deadline with its review, for a deluge of data was pouring in from various studies. FDA extended the deadline.

One report showed in animal studies that high doses of saccharin exerted a co-carcinogenic effect; i.e., while saccharin itself had not initiated cancer, it had promoted the action of other substances that resulted in cancer. The researchers suggested that mixtures of saccharin and other substances, such as cyclamates, would be undesirable.[77] While cyclamates were banned in the United States they were still permitted elsewhere.

Another study, reported in late 1973 by chemists at the National Institutes of Environmental Health Sciences, was the first to test saccharin with animals at average human use levels. Levels were those likely to be consumed by humans, i.e., two tablets of saccharin in a cup of coffee. Saccharin used over long periods of time accumulated in the animals' bladders. The researchers recommended that regular saccharin users should discontinue saccharin consumption occasionally for several days to allow for tissue clearance.[78]

At this point, while some studies appeared to incriminate saccharin, newer studies appeared to exonerate cyclamates.[79] Rumors spread that saccharin might be banned and cyclamates reinstated. FDA faced a dilemma. On one hand, a mistrustful public viewed the agency's actions as overdue, indecisive, and lacking safety assurance. On the other hand, affected economic interests and heavy saccharin users viewed the agency's actions as arbitrary, capricious, precipitous, and unscientific. Whatever the decision, the agency would be faulted.

The lessons of the cyclamate affair made everyone wary of making decisions that might be viewed as hasty or taking action before incontrovertible evidence was available. In evaluating saccharin, one NAS reviewer remarked, "this is a job for a philosopher, not a scientist." He described his efforts as "an education in obtuseness — poor data derived from poor experiments." [80]

For the new saccharin review, NAS's principal task was to examine unpublished reports of research conducted subsequent to the Food Protection Committee's 1970 review. After two and a half years of deliberation, on January 9, 1975, the committee handed FDA the long-awaited report, *Safety of Saccharin and Sodium Saccharin in the Human Diet*. The conclusion was inconclusive: "The results of toxicity studies thus far reported have not established conclusively whether saccharin is or is not carcinogenic when administered orally to test animals." The committee admitted that the results of studies by FDA and WARF suggested that bladder tumors were related to saccharin consumption, but they could not be interpreted as showing that saccharin itself caused the tumors. At the same time, faultily designed tests failed to prove that saccharin is *not* a bladder cancer-inducer. The committee recommended additional studies to resolve the carcinogenic question as well as other safety issues, including saccharin's ability to be transferred from the pregnant woman through the placenta and possibly induce cancer in the developing embryo, the toxicologic significance of saccharin's impurities, changes in urine composition at high saccharin levels and their relationship to bladder stones or calculi, and epidemiologic studies relating cancer incidence with long-term saccharin consumption.[81]

FDA expected to examine NAS's review to determine, in consultation with the committee, what further tests were needed and how they should be conducted. Meanwhile, saccharin would continue under its interim status.[82]

NAS's review was criticized sharply from various sources. FDA's director of Scientific Liaison charged that "the overall tenor of the report leaves the unmistakable impression that a group of saccharin defenders were out to beat back the saccharin accusers no matter what the cost to logic and scientific impartiality." He also disputed the committee's contention that saccharin could not be judged unsafe for humans on the basis of tumors in rat bladders. This was precisely how the judgment on saccharin's use as a food additive must be made.[83]

WARF's director was also critical of NAS's review, noting discrepancies in the committee's interpretations of WARF's findings. Only half of the saccharin-related tumors noted in WARF's study was included in the committee's review. FDA took no action to clear up this discrepancy.

Numerous studies evaluated in NAS's review were from unpublished sources and unavailable for evaluation by others. This feature was criticized by Anita Johnson, staff attorney for a public interest group. In her attempts to examine eight negative studies reviewed by NAS that allegedly disproved saccharin's carcinogenicity, Johnson found that two studies were inaccessible and four were uninformative and essentially useless. Two studies made grossly inadequate animal examinations. One study involved only seven animals. In another, no microscopic evaluation was made of the bladder, although bladder tumors are not always visible to the eye. In one study, control animals had such a high incidence of pituitary tumors that, according to NCI analysis, no chemical could be proven carcinogenic with such controls.[51]

Multi-sites, with various organs as targets for cancer, is a common occurrence. Saccharin may be responsible for cancer in sites other than the bladder. In a preliminary analysis, Johnson found four positive studies. In a mouse study, overall tumors, primarily in the lung, increased significantly in saccharin-fed animals. The study corroborated earlier FDA findings. In a Japanese study, saccharin-fed animals showed a dramatically high incidence of overall tumors, while in a Canadian study, a twofold increase in leukemias and lymphomas resulted.

By the mid-1970s, even persons who wished to shun saccharin consumption were exposed. Saccharin was in illegal use by some food and beverage processors as a partial sugar replacer to offset high sugar prices. FDA and state regulatory agencies were finding bakery products, candies, chocolate milk, frozen desserts, and other products customarily containing large amounts of sugar, adulterated with saccharin. The presence of saccharin was undeclared on labels, or cleverly disguised under the deceptive phrases *flavor* or *flavor enhancer*.[84]

Saccharin continued to enjoy the privilege of interim status for about six years while its safety questions remained unresolved. During those years, at least 23 new studies indicated that saccharin might be carcinogenic. Senator Gaylord Nelson (D.–Wis.) charged that FDA was violating the law by allowing continued saccharin use without a final determination of safety, and he requested the General Accounting Office to review the case. On August 16, 1976, GAO issued a report, *Need to Resolve Safety Questions on Saccharin,* and urged FDA to make a final decision. The continued interim status

> seems contrary to the Food and Drug Administration's intent of permitting use of such additives for a limited time. Extended use of a food additive, such as saccharin, whose safety has not been established and for which a question of carcinogenic potential has been raised could expose the public to unnecessary risk.

GAO expressed concern that the narrow safety margin of only thirtyfold was questionable, in view of many doubts that had been raised about saccharin's cancer-causing potential. GAO recommended that authorized levels of food use should be based on the higher, more generally accepted hundredfold safety margin.

GAO found that the impurity in commercial saccharin, similar in chemical structure to acknowledged carcinogens, was permitted at levels much higher than could be achieved with good manufacturing practices. Originally, FDA had set the tolerance level at 100 parts per million based on industry's capability. But newer technological advancement made it possible to reduce the level to less than 50 ppm or even as low as 1–3 ppm. GAO recommended that the permitted impurity level be reduced to the lowest achievable one.

GAO recommended that FDA "promptly" reassess its justification for saccharin's continued use, terminate its interim status, and issue a final regulation that would either continue or ban its use.[83]

Despite GAO's call for prompt action, FDA decided to follow its original intention of not making any final decision until the completion of more studies in 1977, or possibly 1978. Meanwhile, preliminary results of a Canadian study showed that saccharin's impurity could cause an increased incidence of bladder stones as well as an abnormal increase in cell numbers. On January 6, 1977, FDA took action to restrict the impurity to 25 ppm, although the agency admitted that it lacked any method capable of detecting the impurity at this level.[85]

Ultimately, the smoking gun was produced, with the long-sought conclusive evidence. The Canadian Health Protection Branch had undertaken far-ranging saccharin studies, begun in February 1974. By 1977, one study showed "unequivocal" proof of saccharin's cancer potential. Three out of 50 male rats fed saccharin at 5 percent of their diet developed malignant bladder tumors, while the 50 female rats did not. However, of the offspring of these rats, also fed saccharin, 12 males and 2 females developed bladder tumors. In part of the Canadian test, studies indicated that saccharin, not its impurities, was responsible for the malignant tumors.

On March 7, 1977, FDA obtained preliminary results of the Canadian data. Two days later, FDA and the Canadian agency jointly announced actions that would lead to saccharin bans in both countries. FDA invoked the Delaney Clause.

FDA announced that the Canadian data did not indicate

"an immediate hazard to public health" and the agency did not consider the recall of existing products to be necessary. Manufacturers were encouraged, however, to discontinue saccharin use as soon as possible. The agency announced that it would prepare documents as soon as possible, in 30 days or less, to propose the ban, which would be followed by the customary 60-day period for public comment and reaction.[86]

Reactions were immediate. The Calorie Control Council, representing American and Japanese saccharin manufacturers, soft drink companies, and pharmaceutical firms, called FDA's action "an example of colossal government over-regulation and disregard of science and the needs and wants of consumers." One major packer of saccharin-containing tabletop packets issued a statement calling the proposed ban "an outrageous and harmful action based on flimsy scientific evidence that has no direct bearing on human health. To act on the basis of one questionable experiment creates senseless damage to the public and to a $2 billion-a-year industry involving thousands of jobs." In addition, FDA received countless telephone calls and communications from irate individuals who demanded that saccharin be kept on the market. A spokesman for the American Diabetes Association warned that the ban could have "very grave" effects for the 10 million American diabetics.

Consumer organizations and activists as well as many scientists and physicians supported FDA's action. The Canadian Diabetes Association, the Canadian Medical Association, and surprisingly, even a spokesman for the Canadian soft drink industry supported the Canadian action.[87]

"If society is to make progress in preventing cancer, then the Food and Drug Administration should be commended, not condemned, for banning saccharin," wrote Dr. Charles F. Wurster, an environmental scientist. Near hysterical criticism was directed, "not at the cancer hazard, but at those who would protect us from it and even at the law they upheld,"

Wurster noted. The notion was misleading, he said, that saccharin had been used safely for decades without inflicting human harm. Most cancers are caused by environmental factors, yet only few human carcinogens are identified. The exact cause of the overwhelming majority of cancers remains unknown. Developed tumors do not bear identification tags naming the substances that, decades earlier, induced their cancer development. Saccharin may cause many cases of cancer, yet we have no way to establish the fact with certainty. In the human population, very large numbers of people are exposed to low levels of chemicals, but the impact of seemingly low doses may not be low for a carcinogen. For example, exposure of more than 200 million Americans to doses that cause one cancer in only 10,000 people would result in 20,000 cancers, a number that would be regarded as a public health disaster.[88]

Three misleading statements made by saccharin proponents were repeated endlessly. Even now, these canards are being used, not only regarding saccharin but applied to many other substances.

(1) "Humans would need to drink about 875 diet soft drinks or chew 6,700 saccharin-containing bubble gum sticks daily to equal the saccharin dose given to the test animals." The statement was without any scientific credence. A review of 10 rodent feeding studies by a NCI pathologist showed that doses as low as one-hundredth of one percent saccharin induced cancer in a wide range of experimental conditions. This amount corresponds to about one and a half cans daily, not 875. Also, as Dr. George T. Bryan remarked, "whether you drink one can of artificially sweetened pop or [many] cans, you're at risk from the first drink onward." [89] Dr. Richard Bates, FDA's Commissioner for Science, testified that as little saccharin as is contained in just one can of diet pop could induce cancer in rats.[90]

(2) "Anything can cause cancer if given at sufficiently high

levels." Untrue. High doses of chemicals generally safe may
be toxic at high levels but will not cause tumors. Relatively
few chemicals have been shown to cause cancer, even when
fed at the highest possible doses.

(3) "Small amounts of a chemical are safe for people, even
though large doses cause cancer in animals." Untrue. No safe
threshold dose has been identified for any cancer-causing
chemical. In some instances, humans are far more sensitive
than test animals; in other cases, less sensitive. In the case of
thalidomide, the drug that caused malformed human offspring,
the human was found to be 10 times more sensitive than the
baboon, 20 times more than the monkey, 60 times more than
the rabbit, over 100 times more than the rat, 200 times more
than the armadillo or the dog, and 700 times more than the
cat. Therefore, it is dangerous to underestimate animal data
and argue that extrapolation of results to humans indicates
exceedingly small risks. The risks may be great, and such
faulty underestimations may affect many lives adversely.[91]

Cancer causation by a chemical at *any* dosage in laboratory
animals should always be regarded as a warning signal of
potential hazards to humans. Every human carcinogen is also
carcinogenic in animals, and many such well-known carcinogens as vinyl chloride and diethylstilbestrol were shown first
as carcinogenic in humans. These considerations underlie the
scientific consensus that there is no way to predict a "safe"
level of human exposures from the results of animal carcinogenicity tests.

The absence of cancer signs in one strain of species of test
animal does not prove automatically the substance's safety.
For example, the fact that saccharin failed to induce cancer
in monkeys did not eliminate the substance's potential hazard
to humans as indicated by test results of saccharin-fed rats.
Positive evidence does not nullify negative evidence.

These misleading statements and their unscientific bases
were exploited by saccharin proponents to topple the Delaney
Clause. "We ignore cancer-causation of animals at our peril,"

warned Wurster.[88] The Delaney Clause protects us from this folly, wisely allowing no human discretion based on dosage since there is no valid scientific basis for such discretion. Exposure to any amount of a carcinogen, regardless of how low, is hazardous.

For more than two decades, national and international committees of cancer experts considered and unanimously endorsed the Delaney Clause's scientific basis. In 1973, at a New York Academy of Sciences workshop, attended by more than a hundred leading scientists and lawyers from universities, government, and industry, not only was there overwhelming scientific agreement that the Delaney Clause should be retained and reaffirmed, but strong support was given to its extension, to protect against deliberate introduction of carcinogens into air, water, and the workplace, too.[92]

Criticisms of the Clause emerged almost exclusively from industrial groups and trade associations and their scientific consultants. These groups wrongly charged that the Clause denied opportunity for scientific judgment. The Clause specified that before a food additive could be banned, it must be tested appropriately and have produced cancer. Both requirements involve scientific judgment.

FDA, which never supported the Clause, has found numerous ways to circumvent and weaken it and has invoked it reluctantly and apologetically. FDA's press release concerning the proposed saccharin ban misled the public. The agency implied that the banning action was forced upon FDA by the Clause. In reality, FDA had the power and obligation to ban saccharin even without invoking the Clause, as did the Canadian government, which has no regulation comparable to the Clause. Food additives such as saccharin, no longer on the GRAS list, are covered by general food additive regulations. Indeed, on a later occasion, FDA chose to ban diethylstilbestrol, another carcinogen, without invoking the Delaney Clause.

FDA's press release further misled the public by repeating

the lie about the need for daily consumption of 875 diet soft drinks to induce cancer. The agency's scientists knew better. Furthermore, the press release falsely stated that the agency's actions were based on the Canadian study,[86] which gave rise to the ridicule that FDA's action were based precipitously "at the drop of three rats." Actually, FDA's actions were based on at least a dozen or more prior cancer-positive studies, including its own and one from WARF. Additionally, saccharin was related to a wide range of other cancer sites, particularly in the ovaries and breast; and to different cancer types, such as leukemias, lymphomas, and lymphosarcomas.[93]

On March 11, 1977, FDA announced plans to consider classifying saccharin as a drug so that physicians could prescribe its use for weight control and for diabetes and other disorders for which weight control is vital. Canadian officials announced plans to make saccharin available in pharmacies as of September 1, 1977.[87]

FDA's plans for saccharin's reclassification as a drug was made because of the prospect that with the saccharin ban, no sugar substitute would be available. Despite the concern to retain saccharin for weight control and diabetic use, *the value of artificial sweeteners and artificially sweetened foods has never been proven for these purposes.* On the contrary, animal studies showed that saccharin was an *appetite stimulant* because of its ability to reduce blood sugar.

In 1974, on behalf of the National Institute of Medicine, Dr. Kenneth Melmon had stated that "the data on the efficacy of saccharin or its salts for the treatment of patients with obesity, dental caries, coronary artery disease, or even diabetes has not so far produced a clear picture to us of the usefulness of the drug." [51] This viewpoint was confirmed after the proposed saccharin ban. Saccharin offered no health benefits for diabetics, reported Dr. Harold Rifkin, a diabetician associated with the American Diabetes Association in New York. Nor did saccharin offer any health benefit for dieters, according to Norine Condon, a dietitian with the American Dietetics Association in Chicago. "It's very much a psychological thing," said Condon. "People are used to eating lots of sweet things. They don't want to give up that flavor." [94]

As with news of the earlier ban on cyclamates, news of the impending saccharin ban sent heavy users scurrying to stores and sweeping the shelves clean, to hoard supplies as future hedges. Food and beverage processors ignored FDA's suggestion to discontinue saccharin uses and continued to produce record-breaking volumes of saccharin-containing products.[95]

From the inception of the Delaney Clause in 1958, food and food chemical interests opposed it. When minor substances were banned, opponents grumbled. When bans were of great economic significance, however, such as the 1969 cyclamate ban, and in 1973 with the first diethylstilbestrol ban, opponents mounted stronger attacks. Prompted by the proposed saccharin ban in 1977, opponents declared full-scale war in their efforts to topple the Clause, once and for all.

The Calorie Control Council had a multimillion-dollar war chest and was reported to have spent an estimated $2.5 million in the saccharin battle. Within days of the proposed saccharin ban, the council had placed some 32 full two-page ads in major newspapers throughout the country, questioning the scientific tests, attacking FDA, and urging readers to protest the proposed ban to congressmen. The council distributed materials nationwide to news media and retained a prominent public relations firm to present the council's views. Within the first three weeks following the ban proposal, the council invested $1.14 million on congressional lobbying. "Our basic thrust used to be scientific in nature," the council's chairman said, when interviewed. "Recently, we've had to change our arena of action. We've become political rather than scientific in nature." [96]

Political action proved successful. Ten days after the ban proposal, the U.S. House of Representatives held hearings on the issue.[97] Sessions were stormy, with charges and counter-charges volleying between saccharin defenders and opponents. Throughout the hearings, the Delaney Clause loomed large as a pivotal issue.

Some Washington, D.C., observers believed that the saccharin issue was chosen purposely as a major battleground, not only to discredit FDA but for food interests to gain power in other disputes over even more economically important food additive issues. "This is the best issue that we've ever had to take the FDA to task," reported a vice-president of a leading soft drink company. "If we don't do it now, we never will," he added. "There are 1,800 flavors we use in foods and the FDA will try to attack each and every one . . . Industry cannot live with an FDA that cannot accept science and creates a science of [its] own." [98]

Continued pressure on Congress, largely from the direct action of the Calorie Control Council and indirectly from individuals spurred on by the council to take action, succeeded in making some legislators call for modification of the Delaney Clause to allow saccharin's continued use. Dozens of bills were introduced in the House of Representatives to stop the proposed ban.[99]

Meanwhile, saccharin studies were continued. The House of Representatives' Subcommittee on Health and the Environment, which had held the saccharin hearings, requested NAS and NCI to review the Canadian study so that FDA's evaluation would be double-checked by independent assessment.[100]

Also, Congress financed an additional saccharin assessment by its own Office of Technology Assessment (OTA), scheduled to do short-term testing for mutagenic or other genetic changes. These tests were intended to serve as a basis for evaluating a number of bills pending in Congress dealing with the saccharin issue. Preliminary findings from OTA's panel showed that saccharin was a weak cancer-causing agent, and there was no indication that the substance could be presumed safe for human consumption. While the media treated the term "weak carcinogen" as relatively good news, the truth is that a weak carcinogen is of great concern to scientists. Such substances are not readily detected, and for this reason long and widespread exposures to weak carcinogens are apt to occur. Being more readily identified, strong carcinogens have a better chance of being withdrawn more promptly, and exposures to them are apt to be briefer and less widespread. One OTA panel member noted that if saccharin is implicated in human bladder cancer, indications would be expected to turn up only about the year 2000. Saccharin use began to peak during World War II, when sugar was scarce, and sixty years would be the minimal time cancer epidemiologists could expect the disease to develop after high exposure.[101]

The public outcry against the saccharin ban continued. On April 14, 1977, FDA issued a compromise plan, intended to ease the proposed ban. The agency proposed to permit the continued marketing of saccharin as an over-the-counter non-prescription drug provided that manufacturers could prove its medical value. Saccharin would be withdrawn as a general purpose food additive as well as a non-medical ingredient in drugs and cosmetics likely to be ingested, such as lipstick, toothpaste, and mouthwash. Clearly, as with cyclamates, proof of its medical value might be an obstacle.[102]

FDA's proposals failed to halt the protests. Instead of calming the controversy, the agency's proposals seemed to rekindle the furor.[103] In May 1977, FDA held two days of public hearings. Nothing new was added to the body of scientific knowledge. In a circuslike atmosphere, some demonstrators waved pro-saccharin placards before TV cameras, while others wheeled shopping carts filled with saccharin-containing products.[104] On June 3, 1977, Representative Paul G. Rogers (D.–Fla.) announced that he would introduce a bill that would place an 18-month moratorium on any FDA action with saccharin.[105]

However, on June 18, 1977, FDA announced a new Canadian study that linked saccharin to bladder cancer in men. This study reinforced FDA's position, and the agency officials thought that release of news of this as-yet-unpublished study might win them more support, offering as it did conclusive evidence linking saccharin to human cancer. The Canadian study showed that men who used saccharin or cyclamates (which were still legal in Canada) had a 60 percent higher chance of developing cancer than non-users. FDA considered the new study sufficiently important to merit scrutiny by the scientific community and the public. The agency extended the comment period, which was due to expire on June 14, 1977, and announced that it also would delay its ruling beyond the projected midsummer target date. As events broke, the target date became irrelevant. Regulatory control was about to be snatched away from FDA, as some 315 congressmen agreed to sponsor legislation to head off the proposed saccharin ban.[106]

By early July 1977, both houses of Congress worked on legislation. Shortly before the summer recess, the Senate Health Subcommittee approved a bill placing a moratorium on FDA's ban pending a study by NAS's Institute of Medicine.[107] The House committee approved a similar bill and appropriated $1 million for FDA to permit further tests, with saccharin and cyclamates, and any other sweeteners that FDA might judge in need of more definitive tests.[108]

Shortly after Congress reconvened, on September 15, 1977, the Senate held a day-long debate on saccharin. The prime opponent of the proposed moratorium, Senator Gaylord Nelson, called the Senate bill "a fundamental attack on the best food and drug law in the world." Hale Champion, undersecretary of HEW, the highest government health official involved in the saccharin issue, warned against any delaying legislation. Champion called the moratorium a bad precedent of congressional interference in the regulatory process.[109]

Over this opposition, the Senate voted an 18-month moratorium on any saccharin ban, but required a warning label on saccharin-sweetened products and on vending machines that dispensed such products. Shortly after, the House voted for similar legislation, and on November 23, 1977, signed by President Jimmy Carter, the moratorium went into effect.[110]

By November 1977, the OTA review for Congress was completed. In 12 short-term tests, three were positive, and the OTA panel reported a "clear suggestion" that saccharin was mutagenic. Also, saccharin was carcinogenic in rats and mice. The panel concluded that saccharin was a weak human carcinogen.[101]

As follow-up to the moratorium, saccharin studies were planned to meet the specific requirements of the legislation. Once again, FDA requested NAS to review the scientific literature and assess saccharin's so-called health benefits and risks, as well as the social and economic impacts of its regulations.

FDA's 11-member Saccharin Working Group evaluated available epidemiological data and concluded that they were insufficient to either accept or reject the idea that saccharin use increases human bladder cancer risk. The group proposed a new study that, it hoped, would avoid the pitfalls of past epidemiological studies with saccharin. The study would include sufficient numbers, have controls, choose people from the general population rather than from hospitals, and from selected high- and low-risk areas. Based on these suggestions, FDA and NCI jointly announced on January 25, 1978, a nationwide study involving some 9000 people: 3000 with identified bladder cancer, and 6000 randomly chosen healthy individuals from the same areas.[111] Scheduled to begin in March 1978, the estimated cost of this 18-month study was $1.375 million; by completion, it reached $1.5 million.[112]

On November 4, 1978, NAS released the first half of a two-part review. For the very first time, NAS's committee stated conclusively that saccharin "must be viewed as a potential cause of cancer in humans" not only because it was a weak carcinogen but also because it was a co-carcinogen. The latter fact was possibly even more important than the former. The evidence linking saccharin to cancer in test animals was so strong that further testing was not needed. There was no evidence that saccharin has any health benefits. The committee expressed concern that, based on intake per body weight, the greatest saccharin consumers of any age group were children under 10, mainly with diet drinks. Since cancer may take years to develop, high saccharin consumption over many years placed children at particular risk.[118]

By March 1979, NAS released the second half of its review. Saccharin was only part of the review, which dealt with the broad issue of revamping food safety laws. The committee urged a more flexible food safety policy that would give FDA options other than a complete ban for food additives suspected as carcinogens. However, the committee members were unable to agree about how saccharin should be handled, either under the proposed scheme or by Congress. The committee suggested a plan for assigning categories to food substances, such as high, moderate, or low risks. Some members disagreed as to whether saccharin was a high or moderate risk, and questioned the appropriateness of categorizing carcinogens or other compounds that inflict irreversible damage.[114]

On May 9, 1979, the Senate held hearings to assess saccharin's risks in view of newer findings and to hear arguments for and against extending the moratorium. Among the newer findings was a significant report, based on work conducted by the American Health Foundation in New York. At first, the study appeared to prove saccharin's safety, but by November 1977, results showed the opposite. Bladder cancer risk in men who were saccharin users was nearly twice that of non-users. The findings, sent to NCI at the time, were not released publicly. The timing coincided with the congressional moratorium.

A delegation from NAS's committee took the saccharin issue and its attendant questions, with which the group had been grappling long and hard, and handed them back to Congress. The delegation explained at the Senate Hearings that the saccharin issue would not go away on its own, and that Congress needed to make some decision.[115]

Congress, with three options, faced a no-win situation. It could extend the moratorium, leaving itself open to charges that the economic impact outweighed health risks. Congress could return the issue to FDA, which would begin banning action; and in this case, the saccharin proponents would mount another vigorous campaign to defeat any proposed ban. Or, Congress could attempt to modify the Delaney Clause, in which instance, saccharin opponents, joined by a broad spectrum of groups interested in public safety issues, would coalesce in attempts to keep the Clause intact.[116]

With the nearing of the moratorium expiration on May 23, 1979, congressmen introduced new legislation for its extension. No bill could pass before the imminent deadline, but FDA promised no immediate action.[117] FDA awaited indication of congressional intentions. Even if the agency again proposed a saccharin ban, the process to allow time for comments and reviews would require a minimum of 15 months before any final action. The head of NAS's committee studying saccharin's risks urged consideration of an orderly three-year phaseout of saccharin.[118]

On the day when the moratorium was due to expire, the House began saccharin hearings. Following the hearings, the House passed legislation to extend the moratorium, and later the Senate followed suit. The moratorium was extended until June 30, 1981, a compromise between proposals to extend it for another 18 months or for three years. The extension was intended to give both Congress and the scientific community time to develop a reasonable national policy and to provide FDA with greater flexibility in regulating food additives like saccharin. When the second moratorium expired, Congress extended it for another two years, ending on June 30, 1983.[119]

In late 1979 and early 1980, three epidemiological studies of saccharin's potential role in human bladder cancers were released. The manner in which the media presented these studies led the public to believe that the new evidence refuted the Canadian rat tests by which saccharin had been condemned. SACCHARIN SCARE DEBUNKED and CANCER RISK DENIED FOR SWEETENERS were typical newspaper headlines. The diet food and soft drink industries magnified the distortion and claimed that saccharin safety had been affirmed. Understanding of the issues by Congress and the public was shaped by this false perception.[120]

In reality, the three new epidemiological studies did *not* refute the animal tests. In general, they were compatible. The first one, released on December 20, 1979, was the 18-month study conducted by FDA and NCI, which confirmed earlier findings that saccharin was a human cancer risk. Since patients were questioned about their past as well as present use of artificial sweeteners, cyclamates were included along with saccharin. Other possible carcinogenic factors in lifestyle also were considered. Preliminary findings indicated that there was no increased risk of bladder cancer among artificial sweetener users in the overall population. However, heavy users (six or more servings daily, or two or more diet beverages daily) showed a 60 percent increased risk of bladder cancer. Heavy cigarette smokers (two packs a day for men, more than a pack a day for women) who were heavy artificial sweetener users as well, had a higher risk of bladder cancer than heavy smokers who were not artificial sweetener users. Women, who normally would be at low risk for bladder cancer, but who used artificial sweeteners or diet beverages at least twice a day, were at 60 percent greater risk of cancer than similar women who had never been artificial sweetener users. The risk increased with the amount consumed. The study, though important, was thought not likely to settle the saccharin controversy because the demonstrated risks appeared to be limited to certain types of users. Half the subjects in this study were 67 years or older and, in proportion to their body weight, consumed far less artificial sweeteners than their children and grandchildren. Assessment of the possible effects on young people consuming large amounts over long periods can only be made 20 to 30 years in the future.[121]

The study's prime purpose was to determine if the cancer rate was much higher than expected, as had been suggested by one earlier but questionable epidemiological survey. This study showed that the cancer rate was not much higher than expected. *But this answer was entirely different from being proof of no effect,* which was the interpretation given by the media.

The second study, conducted by the American Health Foundation and published in March 1980, found no statistically significant differences in bladder cancer patients who were saccharin users or non-users.[122]

The third study, conducted by the Harvard School of Public Health and also published in March 1980, showed that men who consumed more than three artificially sweetened soft drinks a day had a greater risk of developing urinary tract cancer than men who consumed less. Increasing the frequency or duration of the use of artificial sweeteners was not associated consistently with increasing relative risk. Although the researchers concluded that "users of artificial sweeteners have little or no excess risk" of lower urinary tract cancer, they admitted that more time may be necessary to accumulate a carcinogenic exposure level.[123]

Congress viewed these three new epidemiological studies as confirmation of previous negative results and its own wisdom in staying the ban. A large segment of the public was relieved to learn that their saccharin consumption was not very risky. Most scientists came to the opposite conclusion. The scientific community recognized that the three new studies needed to be interpreted in conjunction with the well-conducted animal studies. As a group, the animal studies suggested that saccharin exposure increased human bladder cancer risk in a range centering around 4 percent. But even the most elaborate of the three new studies, that made by FDA-NCI, would detect no change in cancer incidence smaller than 15 percent. Even prior to the study, officials at FDA and NCI knew that positive bladder cancer detection would be highly unlikely. The American Health Foundation and Harvard studies included far fewer patients and were even less likely to detect evidence of saccharin's carcinogenicity. Neither study did.[124]

The moratoria provided additional time to gather and study new evidence, but they did not solve the dilemma. Saccharin's ultimate fate may depend on additional safety data, congressional action on revamping food additive regulations, or approval of substitute sweeteners. Meanwhile, the statement by Dr. Robert Hoover, an NCI researcher, which reflects the general medical and scientific consensus about saccharin, seems to be good advice.

> When all the evidence of toxicity is weighed against the lack of objective evidence of benefit, any use by non-diabetic children or pregnant women, heavy use by young women of childbearing age and excessive use by anyone are ill-advised and should be actively discouraged by the medical community.[123]

Rare Sugars:
How Useful? How Safe?

Sorbitol, mannitol, and xylitol

Other alternative sweeteners to saccharin for dietetics are sorbitol, mannitol, and xylitol. They are sweet alcohols which impart a sweet taste, and hypothetically they might ultimately become useful as nutritive sweeteners, according to Annie Galbraith, immediate former president of the American Dietetic Association. However, she pointed out that if such sweeteners as xylitol, sorbitol, mannitol . . . were to be used, "it would be extremely important to consider the calorie content, which is similar to sucrose, and to counsel the patient accordingly." Further study, she said, is needed to establish the usefulness of these products.

— *Food Processing,* May 1977

NOW WE COME to a group of non-glucose carbohydrates called variously "rare sugars," "sugar polyols," "sugar alcohols," "hexitols," "hexahydroxy alcohols," or "hexahydric alcohols."[1] The "hex" prefix is given since rare sugars are six-sided chemical structures. Although some 20 rare sugars have been identified, only three are applied as sweeteners or as carbohydrate sources for intravenous feeding of patients.[2]

To date, rare sugars have been neglected, and information about their safety is woefully inadequate. Currently, FDA is evaluating the sparse existing toxicologic data.

As non-glucose carbohydrates, rare sugars are absorbed from the alimentary tract more slowly and to a lesser degree than common dietary sugars. This characteristic of poor absorption can lead to gastrointestinal problems if rare sugars are consumed at high levels. As with fructose, rare sugars are metabolized somewhat differently from other sugars and possibly may offer some advantage to diabetics. Unlike other sugars, most of the initial metabolism of rare sugars occurs in the liver, independent of insulin. Since they become partly converted to glucose, however, *rare sugars are not entirely independent of insulin.* Rare sugars require less insulin than common dietary sugars during any given time period to keep the blood sugar level constant. Due to this characteristic, they were used originally for special dietary foods intended for diabetics.[3] As long as they were used in limited quantity, they appeared to be safe. However, as the search for sugar alternatives intensified, food and beverage processes turned to rare sugars for widespread applications with products intended for the general public. This extended use has created potential problems.

Sorbitol

How could a research project aimed at finding a new explosive for blasting caps ultimately lead to the development of a market for a sweetener? In 1923, the Atlas Powder Company was searching for a new detonating agent. One promising candidate was mannitol, which was made by sugar electrolysis. However, for every pound of mannitol produced, there were four pounds of a white crystalline powdery waste. To produce mannitol economically, some use needed to be found for the by-product. The powder was slightly sweet and, when investigated further, showed remarkable properties. The powder could control the water content of products with which it was mixed; when esterified, it acted as an emulsifier and surfactant.

The substance was commercial sorbitol, which in time became far more significant than its parent compound. Commercial sorbitol and its derivatives won markets ranging from its use as a food and beverage sweetener, to detergents, pharmaceuticals, and paints.

Natural sorbitol, derived from sorbose, was identified in 1806 by Joseph Louis Proust in the ripe berries of the mountain ash. It is a carbohydrate found in many berries, cherries, plums, apples, seaweeds, algae, and is even detected in blackstrap molasses. Commercial sorbitol is made from dextrose.

In 1929 sorbitol was granted GRAS status as a sweetener for special dietary foods intended for diabetics.[4] Up to 7 percent sorbitol was allowed for this purpose in such foods and beverages.[5]

With increased interest in sugar alternatives in recent times, processors viewed sorbitol for possibilities of other applications. They found sorbitol was versatile. Among its many virtues, sorbitol could promote the retention of original food quality during shipment and storage. It could improve food texture, since sorbitol acts as a crystallization modifier, humectant, softening agent, controller of sweetness or viscosity, and aid in rehydration. For these purposes, processors use sorbitol in baked goods, frostings, and gelatin puddings; frozen dairy products; poultry, fish, meat, and nut products; snack foods; processed fruits; fats and oils; alcoholic and non-alcoholic beverages; and sweet sauces, seasonings, and flavorings. Among sorbitol's special uses, it acts as a release agent in candy manufacture to help slide products out of pans. Compressed bite-size cereal cubes intended for astronauts may be treated with sorbitol to prevent the cubes from crumbling and scattering in a weightless environment. Added to the formulas for cooked sausages and frankfurters, sorbitol helps to remove their casing. It reduces charring and carmelization of processed meats when they are cooked in direct contact with heated metal, for example, grill roller bars. Sorbitol leaves no bitter aftertaste and is used to mask this quality in saccharin-sweetened foods and beverages. Sorbitol also provides body and mouthfeel in low-calorie drinks. As it is being consumed, sorbitol tends to draw heat from the mouth, giving a cool sweet taste sensation.[6]

Some advertising and labeling of sorbitol-sweetened products, especially candies and chewing gums, have been misleading, if not downright deceptive. One sorbitol-sweetened candy proclaims:

> Sugarless candy . . . Free of sugar . . . Low in calories . . . Safe for diabetics . . . Can't hurt children's teeth . . . the secret of [brand name] amazing similarity to sugar-based candy is sorbitol, a safe, natural, nutritive sweetening agent extracted from the skins of fruits and berries. A completely safe, and . . . medically approved substance . . .

"Sugarless" or "sugar-free" are misnomers for sorbitol-sweetened products. Sugarless is *not* free of sugar, but rather free of sucrose. This misleading label term is tolerated by two federal regulatory agencies, FDA and FTC. Some state and local authorities believe that the term is deceptive. The Connecticut Agriculture Experiment Station, which has had an admirable history of testing food products in Connecticut markets, has seized "sugarless" products for mislabeling. Connecticut authorities assert, "The sugarless statement is misleading. Sorbitol and mannitol, declared ingredients, are metabolized as sugars." The Department of Consumer Affairs, in Syracuse, New York, has warned residents of the misrepresentation of "sugarless" products.[7]

"Low in calories" is deceptive. Sorbitol is a carbohydrate with approximately the same number of calories as sucrose. The truth is that sorbitol has only from about 50 to 70 percent the sweetness of sucrose.

"Safe for diabetics" is a claim that needs a qualifier. Sorbitol should *not* be used by diabetics who are untreated or poorly controlled, according to the American Diabetes Association. Also, the types of products sweetened with sorbitol further encourage food choices from nutrient-low foods and beverages, rather than selections from basic nutrient-rich ones.[8]

"Can't hurt children's teeth" is untrue. Dental caries are defined as "localized progressive decay of the teeth, initiated by demineralization of the outer surface of the tooth due to organic acids produced locally by bacteria that ferment deposits of dietary carbohydrates." It is believed that *Streptococcus mutans,* a unique bacterium, is mainly responsible for human dental caries. *S. mutans* can transport and metabolize sorbitol, as well as other carbohydrates. Admittedly, animal tests show that sorbitol is markedly less cariogenic than sucrose, and that dental plaque incubated with sorbitol produces little acid. Few bacteria in plaque are capable of fermenting sorbitol. However, *S. mutans* can grow rapidly and produce acid, incubated in test tubes with sorbitol as the primary carbohydrate source. Consuming sorbitol between meals, which frequently is the case with candy and chewing gum, *can* enhance *S. mutans'* ability to compete with other oral bacteria. *In the absence of other fermentable carbohydrates,* S. mutans *is able to metabolize sorbitol.* Also, there are other reasons why candy and chewing gum are undesirable, apart from the sugar issue.[9]

"Sorbitol . . . extracted from the skins of fruits and berries" is misleading. While it is true that sorbitol is a constituent in fruits and berries, commercial sorbitol is prepared from glucose, a simple, inexpensive, readily available sugar.

"A completely safe and medically approved substance" is unfounded. In addition to sorbitol's use as sweetener, it has been used therapeutically as a carbohydrate source in intravenous feeding. It was discovered that sorbitol can affect the body's ability to absorb and utilize certain nutrients and drugs, either by inhibiting or enhancing them.[10] This feature deserves attention, especially if sorbitol is used at high levels either as a sweetener or as a therapeutic nutrient.

In one medical report, 30 percent of patients who orally

consumed 10 grams of sorbitol suffered reduced absorption of vitamin B_{12} to within the range of pernicious anemia. This condition was inflicted from a single high sorbitol dose and was observed two or three days later. When the dose was increased upward to 50 grams, nearly all patients lost their ability to absorb vitamin B_{12}, and the condition was no longer reversible. Similar inhibition of vitamin B_{12} absorption was demonstrated in experiments with pigs, rats, and guinea pigs.[11]

Also, sorbitol can affect the absorption of vitamin B_6. Sorbitol was found to *stimulate* production of some B vitamin fractions by affecting micro-organisms in the intestinal tract.[12]

At times, sorbitol is added to vitamins and other nutrients in pharmaceutical preparations to increase their absorption. While this absorptive quality may be beneficial for certain nutrients, it may be undesirable for other substances. For example, if a soft drink sweetened with sorbitol contains synthetic colors, flavors, and other food additives of questionable safety, sorbitol's inclusion may ensure a greater absorption of the toxic substances.

Sorbitol's ability to cause diarrhea is well known. In fact, medically, sorbitol is used in large amounts as a cathartic. Sorbitol's relatively slow absorption from the intestine may result in osmotic diarrhea and flatulence if individuals consume high daily amounts (from 30 to 50 grams) or single large doses (20 to 30 grams). The intolerance level varies among individuals.[13]

Numerous reports in medical journals describe cases of diarrhea induced from excessive consumption of sorbitol-sweetened foods and beverages. Commonly, cases concern young children who have overindulged with sorbitol-sweetened candy.[14] In 1966, a pediatrician from the Yale University School of Medicine reported that in the preceding 15 months, 10 children ranging from 20 to 36 months had been treated for diarrhea after having eaten large quantities of sorbitol-sweetened candy. This experience led to further study at Yale.

Two groups of children were chosen. One consisted of five- to six-year-olds, and the other, children from 20 to 36 months of age. Each child was given one package of dietetic mints that contained a total of 9.3 grams of sorbitol. After the mints were consumed, the children and their stools were examined. None of the older children experienced any discomfort or change in bowel habits or stool consistency. The stools, examined 24 hours after the candy had been eaten, contained less than one milligram of sorbitol per gram of wet stool and were otherwise normal. The younger children, however, developed diarrheal stools within two to five hours after eating the candy. The stools were abnormal in sorbitol content, which measured 5 to 20 milligrams per wet stool.[15]

Due to the difference in body size, young children may be at far greater risk than adults eating the same amount of sorbitol. Also, young children do not have the fully developed detoxifying mechanisms of adults.

While medical journals have reported the diarrheal effects of sorbitol in young children, hazards from high levels of sorbitol consumption in adults should also be recognized. In 1967, a physician reported the case of an adult diabetic who developed abdominal distension, gas, and diarrhea after eating sorbitol-sweetened candy.[16]

Many so-called dietetic foods, intended for weight-reduction regimes, contain sorbitol and/or mannitol. In July 1980, an internist reported the case of a 29-year-old healthy man suffering from diarrhea and cramps. In attempting to reduce his weight, the patient had been consuming dietetic foodstuffs, many of which contained sorbitol. Daily, the patient had chewed two packages of sugarless gum, two rolls of sugarless mints, and two dietetic candy bars, which totaled about 50 to 55 grams of sorbitol. The diarrhea abated when the sorbitol-sweetened products were withdrawn. The internist suggested that if physicians were aware of the problem, more sorbitol-induced diarrhea cases would be identified. While diarrhea had been merely uncomfortable for this patient, persons with angina, diabetes, kidney problems, or other illnesses, who consume large quantities of sorbitol, risk serious medical consequences.[17]

FDA, aware of the diarrhea problem from high consumption of sorbitol, suggested as early as 1973 that products which might result in personal consumption of more than 50 grams of sorbitol carry warnings: "Excess consumption may have a laxative effect." Currently, the agency requires this label statement only on food products "whose reasonable foreseeable consumption" may result in a daily ingestion of 50 grams or more of sorbitol. The weakness of this requirement is that no one food or beverage contains high levels. The real problem is the enormous escalation of sorbitol use in the American food supply, especially for those individuals who are heavy users of highly processed foods.

On July 31, 1980, the Center for Science in the Public Interest petitioned FDA to issue a regulation requiring *all* food and beverages containing sorbitol to carry the warning label. The agency took no action, but brought to public attention the current labeling requirement.[18]

The total amount of sorbitol used in foods by 1970 was reported to be about seven times more than in 1960. Current figures are unavailable. Daily average intake figures appear to be unreliable and useless. FASEB, examining sorbitol in the GRAS review, expressed concern about an apparent discrepancy. "Even if all of the 105 million pounds were used in food, the per capita per day average intake of sorbitol would be only 654 mg rather than 30,191 mg given in the [National Research Council] table."[5]

Also, FASEB noted that sorbitol begins to exert a laxative effect at levels that are about twice the estimated average adult intake level, and about equal to the estimated maximum adult intake level. *The average consumption levels for children ages 6 to 11 months, and 12 to 23 months are estimated to be close to, or in excess of a laxative level.* However, since the reported average and maximum intake levels are known to be "generous overestimates" FASEB concluded that "the use of sorbitol in food in the present or reasonably foreseeable amounts poses no problem in this regard." [5]

FASEB expressed concern that

the actual consumption of sorbitol may be considerably higher than average consumption in certain segments of the population. These individuals, for dietary reasons, may select foods containing particularly high levels of sorbitol. Currently available food consumption data do not permit [FASEB] to determine the extent and significance of this problem in regard to sorbitol.[5]

While sorbitol's ability to induce diarrhea is well defined,[18] other possible health effects from high level consumption are not. Both animal experiments and long-range human feeding studies are sparse and inconclusive.

Rats fed sorbitol at high levels (16 percent of the diet; 16 grams per kilo of weight) after one year showed a tendency toward hypercalcemia (excessively high levels of calcium in the blood and other symptoms of abnormality) with the appearance in some animals of bladder concretions and a generalized thickening of the skeleton.[5]

Since sorbitol is used frequently by diabetics, any health effects on this group are of special interest. Rat studies related high levels of sorbitol ingestion to vascular and nervous system complications and to cataract formation — all health problems frequently encountered with diabetes.[19]

Changes in aging may result from lifetime accumulation of sorbitol in cells. Sorbitol penetrates cell membranes poorly. But once it does, sorbitol may become trapped intercellularly and slowly leak. The effect is an accumulation of sorbitol solution inside the cell, which results in increased osmotic pressure. For the diabetic, this is of special concern.

Osmotic pressure may relate to cataract formation, as well as result in damage to the vascular and nervous systems.[20]

In studies at the National Eye Institute, rabbit eye lenses incubated with a high-glucose medium absorbed sorbitol by osmosis. Leakage damaged the fibers, and resulted in lens cataracts.[21]

It must be emphasized, however, that if this result is extrapolated to humans, such possible complications may result from *high levels* of sorbitol consumption over a period of time. The American Diabetes Association's position is that there is no undue need for concern over possible complications of cataracts related to intracellular sorbitol accumulation, but cautions that sorbitol use should be *sparing.*

ADA reported studies with well-regulated maturity-onset diabetes patients, which showed that sorbitol substituted for other sugars, such as sucrose, offered *little if any advantage.* ADA concluded that it appears unnecessary to have specially sweetened foods designed for adults with this type of diabetes. ADA also noted that the effect on diabetics of long-term consumption of large amounts of sorbitol has never been studied.[22]

Sorbitol is being touted as a product "sweetening the foods of the 1980s" in advertisements directed to processors of foods and beverages intended for general consumption. As the number of sorbitol-sweetened products proliferates, the resulting high levels of consumption of this sweetener may be cause for concern.

Mannitol

Mannitol, from mannose sugar, is found in many plants and plant exudates, especially in seaweeds, algae, fungi, and manna ash. The European flowering ash and related plants produce a sweet, dried exudate known as manna sugar that has been used by some primitive Australian tribes as food during famine. Mannitol is poorly utilized by the body and is of limited nutritional value. Commonly, commercial mannitol is made from glucose or invert sugar.[12]

Mannitol is approximately as sweet as dextrose and has the same number of calories as sucrose. Mannitol was given GRAS status for use as a nutrient and/or as a dietary supplement in special dietary foods, with its use limited to up to 5 percent of the food or beverage. In 1963, its use as flavoring or flavoring adjunct was dropped.[23] In 1973, FDA proposed limitations on mannitol's uses, but the National Preservers Association objected and the proposal was never implemented.[24]

Mannitol is used in so-called sugarless candies and chewing gums. It may be used as a release agent in the powdery coating of the gum. While mannitol has limited food uses, it is used in many pharmaceuticals such as breath-fresheners, antacid tablets, cough-cold tablets, and children's aspirin tablets. When consumed, mannitol tends to draw heat from the mouth, which results in a cool sweet taste sensation.

Like sorbitol, mannitol is poorly absorbed. It accumulates water during its slow passage through the intestinal tract and even in relatively small amounts can cause diarrhea. FDA requires a label warning, similar to the one with sorbitol, regarding the likelihood of diarrhea from high levels of mannitol ingestion. While 50 grams of sorbitol may be required, only 20 grams of mannitol is enough to induce diarrhea. The Center for Science in the Public Interest petitioned for warning labels on all mannitol-containing foods and beverages, at the same time it petitioned for similar warning labels on sorbitol. FDA took a similar position with both sugar polyols. The agency brought to public attention the current mannitol labeling requirement.

Traditionally, mannose sugar was widely used in southern Europe as a mild children's laxative. Many immigrants to the United States continued to import it for that purpose.

Mannitol, like sorbitol, is a fermentable carbohydrate in the mouth, although to a lesser extent than sucrose. While few bacteria in plaque can ferment mannitol, *S. mutans* can transport and metabolize mannitol efficiently. In the test tube, when mannitol is the sole carbohydrate source, *S. mutans* grows rapidly and produces acid. Ingestion of mannitol between meals, as with mannitol-containing candy and chewing gum, can enhance *S. mutans'* ability to compete with other oral bacteria. In the absence of other fermentable carbohydrates, *S. mutans* can ferment mannitol.[25]

Mannitol used for intravenous feeding has been associated with a wide range of pathological findings.[26] In 1971, one report noted various manifestations of disturbed metabolism resulting from intravenous infusion with mannitol, ranging from acidosis, dehydration, kidney stones, and kidney failure, to mortality. Some of the conditions were accompanied by symptoms of nausea, confusion, and unconsciousness.[1]

Animal studies showed that a single, repeated, and massive mannitol infusion in dogs resulted in structural and functional changes in the kidneys and brains. Long-term mannitol therapy resulted in serious kidney damage.[27]

In 1979, another danger of mannitol intravenous infusion was reported by a physician who administered repeated high doses (two gm/k every four hours) to a child with Reyes Syndrome, and noted sharply increased osmotic pressure within the patient's skull. If the high dosage of the infusion had been continued, death might have ensued.[28]

As with other rare sugars, safe use of mannitol as a sweetener in the general food supply deserves careful scrutiny.

Xylitol

Xylose, a wood sugar, is widely distributed in plant materials, especially in birch, maple, and cherry wood, as well as in straw, leaves, and hulls. Xylitol, from xylose, can be obtained by boiling corn cobs in acid. Also peanut shells, wheat straw, cottonseed hulls, and coconut shells are practical sources for commercial production. In Finland, a major xylitol producer, birch wood residue provides a plentiful source. After xylitol is extracted, the waste product is used for cattle feed.[29]

Xylitol has about the same sweetness as sucrose and leaves a pleasant cool taste in the mouth. While examining xylitol's potential as a sucrose substitute, Finnish dental researchers made a startling discovery. In 1976, headlines concerning xylitol were ecstatic, from newspapers as widely divergent as *The Wall Street Journal, The National Enquirer,* and *Food Processing.* FINN[ISH] RESEARCHERS HAIL SWEETENER XYLITOL AS "MIRACLE" THAT CAN HEAL TOOTH DECAY, MIRACLE SWEETENER ENDS TOOTH DECAY, and FIVE STICKS A DAY KEEPS THE DENTIST AWAY, were the respective headlines. This enthusiasm was generated by results of the Turku Sugar Studies of Dr. Kauko Makinen, a biochemist, and his colleague, Arje Sheinin, D.D.S., who had divided 125 dental students from the Finnish University of Turku into three feeding groups. The first consumed foods in which xylitol totally replaced sucrose; for the second, fructose totally replaced sucrose; and for the third, sucrose was eaten at an average level. After one year, the xylitol group had 90 percent fewer new cavities than the sucrose group, and 30 to 40 percent fewer cavities than the fructose group.

In another Turku experiment, a hundred students were divided into two groups. One group chewed approximately five sticks of xylitol-sweetened gum daily while the other chewed sucrose-sweetened gum. Both groups were given free choice of food selections. After a year, the xylitol group showed an average *decrease* of one cavity per person; the sucrose group, an average *increase* of three cavities per person. In some students, the process of tooth decay appeared to be reversed, which suggested a possible therapeutic effect from xylitol.

The study demonstrated the role of fermentable carbohy-drates in human caries formation and the fact that *oral bac-teria cannot ferment xylitol.* Apparently, the breakdown of carbohydrates to acids in the mouth is arrested, partly because xylitol itself does not break down in the mouth but does so in the stomach. Also, xylitol raises the plaque pH above the acid range for prolonged periods, inhibits the action of *S. mutans,* and thus retards cavity formation.

In press meetings, Makinen's statements were uncharacter-istic of most scientific researchers. "It's miraculous," Makinen was quoted as saying. He claimed that xylitol "goes beyond" other non-cariogenic sweeteners that supposedly do not pro-mote decay. Xylitol actually prevents it and will even heal incipient cavities.

The "miracle" sweetener generated enormous interest. The segment of American industry that would benefit most was an annual $3.5 billion market of chewing gums, candies, and chocolates.[30]

The National Institute of Dental Health (NIDH) an-nounced that it would launch a three-year study with about a thousand American schoolchildren to confirm the Turku findings. Both American Chicle and Life Savers decided cau-tiously to hold back and await the outcome of the study, while Wrigley decided to introduce a xylitol-sweetened gum. U.S. law did not allow advertisement of xylitol's cavity-fighting ability until completion of the study. Wrigley made no dental claims, but merely described xylitol's sweetening ability and its cooling effect in the mouth.[31]

In November 1977, NIDH's study was launched. Sticks of xylitol-sweetened chewing gum were distributed to school-children on Long Island. The gum was 50 percent xylitol, a level five times higher than in the Wrigley product. The chil-dren were instructed to chew three sticks daily, for a period of three years.

After day three, with only nine sticks chewed, parents were

notified to return the remaining gum supplies. The experiment was halted unexpectedly and abruptly.

This action was triggered by news from the Huntingdon Research Center in Great Britain, where ongoing animal safety studies were conducted for xylitol, funded by Hoffmann–La Roche, the American distributor of imported xylitol. Preliminary results from a two-year mouse study showed that a substantial number of males fed high doses (10 to 20 percent levels) of xylitol developed urinary bladder stones late in life. Of these, some developed inflamed bladders and bladder tumors. In a two-year rat study, with animals fed at similar levels, no bladder stones or tumors developed. But some male rats, fed at the 20 percent level, developed growths on their adrenal glands; in some cases, malignant tumors developed in the adrenal glands.

FDA announced a review of the British findings but took no action to revoke xylitol's food uses. Although some manufacturers, such as Wrigley and others, voluntarily withdrew xylitol from their American products, other companies continue to promote it.[32]

The reality should have sobered health professionals, had they given any thought to the implications of the study. An anti-cavity substance would be carried mainly in a medium that should be used sparingly, if at all. Instead, with a new image of chewing gum as a prophylactic food, it would appear virtuous to chew gum frequently.

Xylitol's use as a *partial* sucrose replacer may not reduce tooth decay. Yet xylitol as a total sucrose replacer in the typical American diet is neither practical nor desirable. Xylitol production is minuscule and costly compared to sucrose. Xylitol's mouthfeel in processed foods and beverages is not identical to that of sucrose. As with sorbitol and mannitol, xylitol, too, if consumed at high levels, will induce osmotic diarrhea.[33]

Scientists have reported that when more than 10 percent

sucrose was replaced with xylitol in rat studies, liver-cell metabolism was damaged. Replacement of 20 percent sucrose with xylitol caused decreased liver glycogen and lipid levels, depleted liver RNA levels, and decreased liver cell growth.

In retrospect, it is shocking to realize that NIDH would launch long-range tests with young children before having carefully and systematically reviewed worldwide literature and research on xylitol. Obviously, if the Huntingdon Research Center had two-year studies in progress, adverse effects were being observed, noted, and discussed. Also, these long-range tests with children were launched with no regard to the diarrheal characteristic of xylitol. This feature of rare sugars was well recognized in medical literature, and it was established that children are especially at risk. In an eagerness to announce a "miracle" sweetener, were the warning signals ignored? To food and beverage processors, xylitol appeared to be one bright hope, being the sole alternative to saccharin, at a time when saccharin appeared doomed. What were the un-heeded warning signals, even before the Huntingdon Research Center news?

In 1963, xylitol was given limited clearance as a sweetener by FDA. Its use was restricted to special dietary purposes as a sugar substitute for diabetics, in gums, jams, jellies, and chewable vitamin coatings.[34] Unlike sorbitol and mannitol, xylitol was not granted GRAS status. Hence, it is legally classified as a food additive and regulated as such.

The first warning signal concerning xylitol came in the early 1970s. In Great Britain, a water solution of 20 percent xylitol became available for patients being fed intravenously who required additional calories. *The Lancet* carried a report of adverse effects suffered by patients infused with xylitol. Of 23, eight developed metabolic acidosis, severe enough so that measures needed to be taken to restore normal blood balance. Seven patients suffered osmotic diuresis, and later four of them developed serious kidney problems that included swell-ings, diminished urine secretion relative to fluid intake, and calcium oxalate crystal deposits in the organ. Effects ranged from mild nausea, to mental confusion, stupor, and complete loss of consciousness. Subsequently, six patients died. At autopsy, unidentified crystal deposits were found in the brains. In Great Britain, medical use of intravenous xylitol infusions was halted. Patients suffered similar adverse effects in New Zealand and Australia, where both oral and intravenous xylitol uses were banned.[35]

After these findings were reported, FDA favored a total ban on xylitol. However, since there had been no indication of harm from low levels of xylitol consumption, FDA favored limiting xylitol's use to chewing gum. The agency was just about to make this proposal when the Huntingdon results were reported. But Hoffmann–La Roche requested that xylitol's uses not be further restricted, since the company was interested in having xylitol available for use in jams, jellies, candies, and other products. FDA acquiesced and never imposed further limitations.[36]

Although intravenous xylitol infusions had been prohibited elsewhere, the practice continued in the United States. In 1976, an American medical report noted similar adverse effects already reported in the literature, and a few additional ones encountered after xylitol infusions were given to surgery patients. New findings included liver injury and calcium oxalate crystal deposits in the artery walls of the midbrain.[37]

Animal studies provided further confirmation. In one study, xylose was found lethal to rats, even within a few days, when fed at normal dietary levels.[1]

In another study, xylitol was examined for its ability to form oxalate crystals in normal rats and in rats deficient in some B vitamin fractions. Oxalate formation increased significantly in B_6-deficient rats infused with xylitol. The study suggested that B_6 deficiency may be a factor contributing to oxalate crystal deposits observed in some patients infused with xylitol.[38]

During the Turku studies, some Finnish students who

chewed xylitol gum and ate large amounts of xylitol in prepared food suffered from diarrhea. This side effect was noted
also in 1977 by a researcher at National Institute for Dental
Research (a division of the National Institutes of Health),
who reported that, in some persons, even moderate xylitol
doses have a cathartic effect. A single diarrheagenic dose is
usually from 30 to 40 grams. However, large individual differences exist in xylitol tolerance, and its sustained use may
lead to adaptation so that higher doses may be tolerated.[8]

Subsequent to that fiasco, additional adverse findings were
reported. In 1979, rats fed xylitol at high levels developed
severe diarrhea and gas. This experiment, conducted by Dr.
Marleen Wekell, a nutritional science professor at the University of Washington, demonstrated how a substance such as
xylitol can adversely alter the bacterial composition of an
organism's intestinal tract. Symptoms of diarrhea and gas were
attributed to the prolific growth of *Clostridium perfringens,* a
disease-causing bacterium normally found only at very low
levels in healthy rats. *C. perfringens* causes similar intestinal
distress in humans and is one of the bacterial species implicated in metabolizing harmless substances into carcinogenic
ones. Since the formation of some carcinogens may depend on
the gut's bacterial environment, this finding underscored the
critical role of diet in health.[39]

At present, xylitol chewing gum is marketed in the United
States. In addition, xylitol is being marketed as a crystalline
sugar. In one advertisement, it was suggested that

SUGAR IS GOOD FOR YOU, IF YOU USE THE RIGHT KIND . . . not
all sugars are alike. The sugar that you were warned against
is over-refined and adulterated cane sugar. Fortunately, there
is an alternative, a natural sugar that provides the delightfully
sweet taste of cane sugar without the side-effects. This sugar
is xylitol — the delicious fruit sugar that is everything sugar
should be — nutritious, energizing and healthful . . . [brand
name] is made from 100 percent pure crystalline xylitol . . .

good for dieters. The reduced insulin secretion works to reduce the appetite. And tests indicate this sweetener may decrease fat accumulation in tissues thereby adding weight loss ... It's the completely natural sugar that's actually good for you.[40]

Despite the hype, xylitol is as refined as cane sugar, and has approximately the same number of calories and sweetening power. Although the advertisement mentions "fruit sugar" most likely the product has been derived from wood sugar. The purported "tests" were neither cited, nor supplied upon request. And the price? $8.50 for a half-pound bottle, which totals *$17.00 per pound*. The saving grace is that at $17.00 a pound, the buyer may use this sweetener parsimoniously.

❉

The use of these rare sugars may be safe when they are consumed infrequently, or at low levels, for dietetic use. However, this former limited application has changed radically. Food and beverage processors, constantly searching for sugar alternatives, now seek new applications for these rare sugars, which are being used increasingly in numerous products intended for general consumption. FASEB noted that, by 1970, the total amount of sorbitol used in foods was approximately seven times greater than the amount used in 1960; and the total amount of mannitol, about ninety times greater than in 1960. This escalation continued throughout the 1970s. The trend is undesirable. Consumption of rare sugars needs to be limited and carefully monitored.

Future Sweeteners:
How Promising?

*Licorice, miracle berry, serendipity berry,
dihydrochalcones, katemfi, stevioside, osladin,
amino acids, synthetic sweeteners,
non-absorbable leashed polymers, non-digestible sugar*

In time, I am confident that an artificial sweetener meeting all the
criteria of safety, stability, solubility, and taste, will be found.
Ideally, it will have the taste of sucrose and the pharmacology of
pure mountain water.
— Marvin K. Cook, consulting chemist, "Natural and Synthetic Sweeteners," *Drug and Cosmetic Industry,* September 1975

Any new substance, despite all the safety tests, is surrounded by
question marks.
— Stanley Glassner, Food and Color Additives Division, FDA, *The
National Observer,* January 4, 1975

USING VARIOUS APPROACHES, the search for alternative
sweeteners continues. All approaches have drawbacks. The
safety of some substances is doubtful, while others remain at
the earliest stages of development. All efforts to develop alternative sweeteners are misplaced. We would be benefited if all
these efforts were scrapped. We do not need alternative sweeteners. Rather, we need to move in the direction of better food
selections, using foods that do not require added sweeteners.

Licorice

Say "licorice" and the response may be "candy." But licorice is used in many more products. Licorice has had a venerable history of use as a flavoring agent and flavoring enhancer in many foods and beverages, pharmaceuticals, tobacco, and snuff, but it had not been used as a primary sweetener. Yet licorice is the sweetest substance with GRAS status, and its derivatives share this quality. Its potential as a natural sweetener has only been considered in the wake of the cyclamate ban and the pending saccharin threat. After the cyclamate ban, the largest licorice producer enjoyed a twentyfold increase of orders, ranging from salad dressing and toothpaste manufacturers to bakers and processors of cured ham and bacon.[1]

Licorice is derived from the root of a plant found mainly in southern Europe and Central Asia. The plant needs to grow four of five years before it is hand harvested. The root, extract, powder, glycyrrhiza, and ammoniated glycyrrhizin are various forms used for flavoring and flavor enhancing.

Food processors found that licorice intensifies other flavors, such as maple, vanilla, anise, root beer, rum, walnut, butterscotch, chocolate, honey, and pickle spices. By adding some powdered or liquid extract of licorice root, they could reduce the amount of costly ingredients. For example, a relatively small amount of licorice greatly intensifies the chocolate character in a product; in some chocolate-based products, licorice may replace up to 20 percent of costly cocoa. Another money-saving feature of licorice is its ability to mellow a general harshness of synthetic flavors.

With foaming and emulsifying action in water, licorice is added to cake mixes, ice cream, ices, candies, baked goods, gelatin desserts, meat sauces, seasonings, and fruit and vegetable products. Although licorice foams, it does not ferment, which makes it useful in dentifrices.[2]

Universally, licorice extract is used to add both flavor and sweetness to pipe, cigarette, and chewing tobaccos. It also helps keep tobaccos moist.[1]

Licorice is used in many pharmaceuticals as flavorant, flavor enhancer, and debittering agent. It may replace saccharin in low-residue diet preparations, breath sprays, mouthwashes, laxatives, and chewable vitamins.

Licorice is 50 times sweeter than sucrose and is synergistic. Licorice, combined with sucrose, yields a perceived sweetness a hundred times greater than that of sucrose by itself.[3] Ammoniated glycyrrhizin, a relatively new derivative of glycyrrhizic acid, an active principle in licorice, exerts an even greater synergistic effect. One pound of ammoniated glycyrrhizin, combined with 100 pounds of surcose, yields a sweetness level of 200 pounds of sucrose.[1]

Ammoniated glycyrrhizin is now processed, by repeated stages of crystallization, until it is converted into a crystalline colorless substance. This significant development turned processors' attention to the potential use as sweetener and sweetener intensifier, which were unanticipated new roles.[3]

As part of the GRAS review, FASEB evaluated the existing literature on the safety of licorice and its derivatives. The panel concluded that the compounds were safe when used as food flavorings at present levels and should be affirmed as GRAS with specific limitations on their use. On August 2, 1977, FDA issued a proposed rule in the *Federal Register* that would affirm the GRAS status for licorice compounds used as flavoring agents, with specific maximum usage levels for various food categories. FDA's proposal was based on the assumption that the sole applications for licorice compounds would be for flavoring and flavor enhancing, which would make the use level self-limiting once the desired effect was achieved.[4]

This was not the case. Increasingly, food and beverage processors were using licorice compounds as sweeteners and as sweetening synergists with sucrose or saccharin, in addition to the traditional uses as flavor enhancer with non-licorice–flavored foods and beverages.

For example, in 1979, possibly as a hedge against future action against saccharin, Alberto-Culver Company test-marketed a saccharin-free product, described as a low-calorie granulated sugar replacement. The available carbohydrate was given as 87.3 percent, and the company admitted that this total included not only the dextrose, which was listed on the panel, but also the "natural flavoring." The sweetener used was ammoniated glycyrrhizin, but not listed by that name on the package. It was hidden as natural flavoring.[1]

The unanticipated expanded use of licorice compounds as sweeteners and sweetening synergists made any self-limiting feature ineffective. FDA recognized that the food use level was outstripping the assumed limitations of the 1977 proposal. In 1979, FDA proposed an amendment to licorice's GRAS status, to limit uses.[5]

The curbing action was prudent. Glycyrrhizic acid, an active principle in licorice, is chemically and structurally related to two adrenal gland hormones (desoxycorticosterone and aldosterone). Glycyrrhizic acid can produce steroidal effects, estrogenic activity, and other biological actions. High licorice consumption may interfere with drugs, by decreasing the efficacy of some used for heart conditions and blood pressure.[6] At times, licorice extract is used in ulcer therapy. Reported side effects include cardiac asthma and swellings in 20 percent of the treated patients.[7]

The safety margin established for licorice use may be inadequate for a segment of the population. Limitations are based on *average* usage and fail to include persons with idiosyncratic dietary habits. Licorice "freaks" are not uncommon, and from time to time, case histories have been reported in medical journals describing the adverse health effects from high licorice consumption. Some of these reports predate the era of increased licorice use as a sweetener and sweetener synergist.

One case, reported in 1968, was a 58-year-old man who daily, for seven years, had eaten two to three licorice candy bars. He developed high blood pressure and paralysis of his

extremities. The syndrome disappeared after licorice was eliminated from his diet.[8]

Another case, reported in 1970, concerned a 53-year-old
man who had eaten about 1.5 pounds of licorice candy (about
700 grams) over a nine-day period. He suffered shortness of
breath, ankle and abdomen swellings, weight gain, headaches,
and weakness. Although previously he had an excellent health
record, his respiratory distress required hospitalization for a
developing heart condition. The physician who reported this
case felt that the medical community should be aware that
overindulgence in licorice-containing products may induce
congestive heart failure.[9]

A 1971 report was about an elderly man who, after moderate exertion, suddenly developed heart palpitations of such
severity as to require hospitalization. He was found to suffer
from hypertension and malfunctioning of the intestine, heart,
and kidney. The patient mentioned that he had been attempting to stop smoking cigarettes, and for the three months prior
to his hospital admission he had been eating licorice candy
daily. It was estimated that he had been consuming approximately 124 grams (nearly 4½ oz.) daily.[10]

A 69-year-old woman was admitted to a hospital in 1980
due to increased muscular weakness. She was a long-time
licorice user, and for the past three months had been drinking
four tablespoons daily of Lydia Pinkham's Compound. This
elixir contains licorice, and it was estimated that the woman
was consuming about 59 ml (12 teaspoons) of licorice daily.[11]

Also in 1980, an 85-year-old man was admitted to a hospital, suffering from a progressive generalized weakness, and
inability to rise from a sitting position for 10 days. He could
not raise his arms above a horizontal position, had moderately
swollen ankles, and profound muscular weakness. The man
had chewed 8 to 12 ounces of tobacco daily, and swallowed
the saliva. The tobacco that the man chewed contained licorice. He had engaged in this practice for half a century. It was

estimated that he had been consuming from 0.88 to 1.33 grams of glycyrrhizic acid daily, within a range known to produce health problems including muscular weakness, hypertension, and potassium deficiency. The potassium deficiency in this man was further aggravated by his low potassium diet. Being toothless, he had been eating canned soup and soft vegetables for more than a year as the sole dietary choices.[12]

In western Europe, a popular alcoholic licorice-containing beverage is Un Boisson de Coco. Persons habitually drinking one to three liters daily have been reported to show muscular weakness, paralysis, tetany (muscular twitchings and cramps), and hypokalemia (low blood potassium level).[13]

Obviously, licorice use in the food supply needs careful monitoring and limitation. Any increased use of licorice compounds, especially to serve as sweeteners to replace other sweeteners, is fraught with danger.

Sweet Perception from Taste-altering Substances

Numerous plants have been found to contain substances hailed as "nature's own sugar substitutes." At first glance, such substances appear to be promising alternatives to synthetic sweeteners. But problems exist. At present, none is permitted for use in the United States, nor are any even close to being marketed here in the near future. Their safety needs to be established before approval can be granted, and that goal has been demonstrated as time-consuming, costly, and frustrating.

Some of these substances are not intrinsically sweet but rather alter taste perceptions by acting on the tongue's taste receptors. Tasting is done through some 7,000 to 10,000 microscopic-size pores, usually clustered on certain tongue papillae, but also found in the upper part of the throat, the rear part of the cheeks, and on the palate.[14]

In the presence of a taste-altering substance, a lemon may taste sweet, or green strawberries may taste ripe. A common experience of altered taste perception is noticed after eating globe artichoke. A sweet taste lingers in the mouth and the sweetness is imparted to anything subsequently consumed. Indeed, some gourmets exclude artichoke from their menus to avoid ruining the taste of fine wine. On average, as little as one-fourth of an artichoke heart causes a taste change comparable to the addition of two teaspoons of table sugar to a six-ounce glass of water. Two chemicals from artichoke, cynarin and chlorogenic acid, have been identified as the main agents for altering taste perception in this vegetable.

The artichoke is only one of numerous plants that contain taste-altering substances. Some plants are from other areas of the world. Many such plants have been investigated for the possibility of developing a new approach to sweetening food. Instead of adding sweeteners, it might be possible merely to create a sweet perception. Furthermore, many of these substances are protein, not carbohydrates, so they would not be cariogenic with the teeth.[15]

Other substances in plants have been found to modify sweet taste perception in other ways. Gymnemic acid, extracted from a tree-climbing tropical plant, is a compound that blocks the sweetness of saccharin or cyclamate for hours, but does not repress other sweet tastes, such as chloroform. The mechanism responsible for this effect is not yet known but, in time, may help us better understand the mystery of sweet taste perception.[16]

Merely because such substances are derived from natural sources does not necessarily guarantee safety. Many plants contain naturally occurring toxins, and toxicological data are sparse. Nor does traditional long-time use automatically assure safety. This assumption has been false in some instances. Three food flavorings derived from natural sources — coumarin from the tonka bean, safrole from sassafras, and oil of calamus from sweet flag — ultimately had to be banned because they were found to be harmful. Agene (a flour bleach) was used for more than 30 years, and dulcin (an artificial sweetener) for more than half a century, before their harm was established.

For these reasons, FDA requires long-range safety tests. This wise precautionary measure, which offers public protection, is not afforded in many other countries. Frequently, it is attacked by those who wish to market products and naively assume that, because a substance is naturally derived, safety tests should be unnecessary. Others, eager to market products, may wrongly anticipate easy and rapid test clearance. Safety tests are extremely costly and time consuming. Some products are never cleared, as time and money run out. The case of the miracle berry serves as an illustration.

The Miracle Berry

The berry from the "miracle fruit" makes sour foods taste sweet. Bushes bearing these red berries grow abundantly in the inland regions of the African Gold Coast. For centuries, the natives have been eating the berries to improve the taste of stale staple gruels turned sour, and to sweeten sour palm wine.

From time to time, during the last three centuries, explorers made references to this remarkable berry. In 1852, W. F. Daniell, a missionary, observed the berry's taste-altering effect and termed it a "miraculous berry." The term held, and the berry is now known as the miracle berry. Daniell took samples back to England, where they were considered a curiosity.

A century later, with the serious quest for alternative sweeteners, this berry was no longer regarded as a mere curiosity. Researchers isolated a protein called miraculin from the berry,[17] which was found to taste 2500 times sweeter than sucrose and capable of altering taste perception for up to 12 hours. Miraculin looked promising.

In 1968, a small company with strong backing and funding was formed in the United States to develop miraculin. The company planted a million miracle berry bushes, developed a process to isolate miraculin from the berry, and successfully completed test marketing. But they failed to anticipate the obstacles ahead.[18]

In 1973, the company petitioned FDA for GRAS status for miraculin. The agency denied the petition and classified the substance as a food additive. That decision signaled the cessation of all marketing. The company was unable to raise additional money to appeal and was unable to fund comprehensive, costly tests necessary to prove safety as a food additive.

In 1974, the company declared bankruptcy. Three years later, FDA reviewed the case and concluded that the data did not "assure the safety of either general or limited use of the miracle fruit and its products for use in foods." [19]

The Serendipity Berry

Another red berry, also from West Africa and known as the Nigerian berry, has long been used to sweeten sour foods in that area. The sweet taste from the fresh berries persists for an hour or longer. Like miraculin, from the miracle berry, the Nigerian berry also contains a soluble, low-calorie protein.[20] Its sweetening ability is, to date, the most intense of all natural substances, being about 3000 times sweeter than sucrose.

The Nigerian berry was rechristened the "serendipity berry" because the fortunate discovery of the source of its sweetening characteristic was made only on the very last day of scheduled research. Robert H. Cagan, Ph.D., a biochemist, was the principal investigator of this berry, and his preliminary studies suggested that the sweet substance in the berry might be a large protein molecule. Some of his work then was furthered by James A. Morris, Ph.D., a postdoctoral student. Morris spent a year purifying and identifying the molecule, which is perhaps 30 times larger than a sucrose molecule. Cagan and Morris named the protein "monellin" after the Monell Center, the only research center in the world devoted entirely to basic multidisciplinary research in the field of chemical senses. Work with monellin at the Monell Center has been geared to this interest. A protein molecule of monellin's size has an interesting possibility for taste research, to unravel the chemical structure required for sweetness.[21]

As to practical application for foods and beverages, industry researchers are not overly enthusiastic. No one knows if the wild berry can be cultivated. Isolation of monellin from the berry is difficult, and commercial production would be expensive. The resulting product is very unstable, and the protein loses its sweetness quite rapidly, especially when exposed to heat or a low pH. The main obstacle, however, in bringing the sweetener closer to the commercial stage "appears to be the regulatory climate that inhibits investment in the studies needed to establish required safety data." [22] The miracle berry disaster is fresh in memory.

The Dihydrochalcones (DHC)

It all started with citrus peel waste. In 1958, scientists at USDA's Fruit and Vegetable Chemical Laboratory at Pasadena, California, were searching for new uses for food industry wastes. The rinds of discarded citrus were used commonly as animal feed. Could other uses be found? Indeed, an unanticipated valuable by-product could be extracted from citrus rind that was discarded from fruit processed for frozen juice or canned segments.

In examining the bitter flavonone glycosides of citrus fruit rind, the scientists made a startling discovery. Naringin, an intensely bitter natural substance present in grapefruit rind, when synthesized into neohesperidin, converts to a substance that is about 1500 times sweeter than sucrose.

To date, unfortunately the only food found to contain neohesperidin as a constituent is in the rind of one variety of orange, the Seville. This variety is common in Spain, but not in the United States. And Spain was not about to export Seville oranges, which are prized as the main ingredient of a world-famous marmalade.

The scientists met the challenge. In several steps, they converted naringin, obtained from plentiful grapefruit rind, into neohesperidin. About 100 pounds of grapefruit rind are needed to make one pound of naringin. But this small amount converts into about 13 pounds of neohesperidin, which has the sweetening power of about 450 pounds of sucrose! [23]

The intense sweetness of neohesperidin has certain drawbacks. The perceived sweetness lags somewhat in starting, but then lasts for a long time, with an even later effect described as a "menthol-cooling" or "licorice-like" sensation in the back of the mouth. These features would be problems for food and beverage use, but possible advantages for products such as chewing gum, candy, dentifrice, and mouthwash. Neohesperidin is approved for use in such products in some countries.

Investigations were extended to other members of dihydrochalcones (DHC). For example, hesperidin dihydrochalcone glucoside is present in sweet oranges. Variations exist among DHC in terms of intensity, time lag, duration, mouth site affected, and their background flavor. It is believed that DHC have potential uses as sweeteners, flavorants, and flavor enhancers.

DHC have undergone extensive toxicologic tests by USDA and commercial producers. While most of the media statements report no apparent harmful effects, some adverse effects have been noted. Long-term studies of animals fed high doses of neohesperidin DHC showed no increase in tumor incidence, but an apparent interaction between the sweetener and the laboratory diet resulted in decreased rat growth. Some dogs showed mild thyroid hypertrophy, elevated liver weights, and testicular atrophy. While it may be argued that such adverse effects were shown only at high feeding levels that may be unrealistic in terms of human consumption, the use of high feeding levels is accepted protocol in toxicologic studies in order to overcome the limitations of animal studies.[24]

To date, FDA has not approved petitions to market DHC.

Katemfi

An African fruit, katemfi, called the "miraculous fruit of the Sudan," contains three large seeds surrounded by transparent jelly. The jelly is intensely sweet, with a slight licorice-menthol flavor. As early as 1839, the coated seeds were traded in West Africa to sweeten bread, fruits, wine, and tea.

In 1972, proteins responsible for the sweetness in the jelly were isolated, identified, and called thaumatin I and II. These proteins are large molecules. By weight, thaumatin proteins are 1600 times sweeter than sucrose. Due to their intense sweetness they could be used sparingly and thus contribute few calories. However, thaumatin I and II are far from any stage of commercial production. As yet, they have not undergone safety tests. Also, the proteins are sensitive to high temperature and low pH, which would limit their potential uses in foods and beverages.[25]

Stevioside

The roots and leaves of a wild shrub grown in Paraguay, Brazil, and Argentina have been used by Indians to sweeten tea and other bitter foods. The Guarini Indians call it "sweet tea." Stevioside, a glycoside isolated from this shrub, has been found to be about 300 times sweeter than sucrose.[26] At a low level, it has an agreeable flavor; at a high level, it leaves a bitter aftertaste. The sweetener is not utilized in human metabolism.[27]

The shrub is cultivated elsewhere, notably in Japan. Since stevioside is from a natural source, it is not subject to safety tests in Japan, where it is in limited use as a low-calorie alternative sweetener.

However, experiments have shown that stevioside is toxic to rats. Moreover, in Paraguay, stevioside is used as an anti-fertility substance, which indicates that this sweetener has biological activity that deserves to be monitored carefully. FDA would require long-term safety studies before approval could be granted in the United States.

Osladin

Osladin is derived from the rhizome of a fern that grows in many areas of the world. Yugoslavian researchers discovered a sweet glycoside extract in the plant, that bears a structural relationship to DHC. Osladin resembles saccharin in sweetness level, which is about 300 times sweeter than that of sucrose.

The yield of osladin from the fern is so meager that its commercial utilization is doubtful. Nevertheless, the fern holds interest. Some compounds related to osladin are being synthesized at the Northern Regional Research Laboratory, USDA, in attempts to relate the sweet-taste phenomenon to certain types of structures.[28]

Amino Acids as Sweeteners

Two unnatural forms of amino acids, D-tryptophan and D-phenylalanine, have been suggested as potential sweeteners. (L- is the natural form for both.) A chlorinated derivative of tryptophan, D-6-chlorotryptophan, was discovered accidentally by pharmaceutical chemists while searching for an anti-fungal enzyme. This substance, about 1300 times sweeter than sucrose, leaves no perceptible aftertaste. Very few toxicologic tests have been performed on these sweeteners, probably because they show less promise for commercial applications than some other substances under investigation. No work on them is reported in progress.

Kynurenine, an amino acid produced in the body by trypto-phan, has been found to be 30 to 50 times sweeter than su-crose. This metabolite is used as N-acetyl and N-formyl derivatives and is reported to be under study.

Glycine, another amino acid, was used with saccharin to mask its bitterness, until FDA prohibited this use. Glycine had GRAS status, which it lost after its safety as a food ad-ditive appeared dubious. Rat experiments showed that glycine, at high levels, suppressed growth, led to body weight loss, in-creased liver weight, and in some cases, paralyzed the cervix.[29]

The aspartame affair may make researchers hesitant to in-vestigate and develop amino acids as sweeteners.

Synthetic Sweeteners

Acetosulfam, also known as A-Silfam-K, is a cyclic sulfona-mide, related to saccharin. Presently, this sweetener is under-going toxicologist tests by Hoechst in West Germany. Results of the studies are not expected to be completed until the mid-1980s, and meanwhile this sweetener is not used anywhere. The taste is similar to that of saccharin, with only half its sweetening power; it is about 130 times sweeter than sucrose. A thiopene analog of saccharin, also under investigation, is about 1000 times sweeter than sucrose. Both acetosulfam and the thiopene analog are stable, like saccharin, under similar conditions of use. However, both are years away from any approval or use in the United States.[30]

Stanford Research Institute (SRI) has been developing a synthetic aldoxime sweetener derived from petrochemicals. Using the acronym of the institute, the product is called SRI oxime V. The sweetener is a modified version of perillartin, which is an intensely sweet derivative of perillaldehyde, an aldoxime non-nutritive sweetener used in Japan since 1920.[31]

At SRI, the testing and development program is in progress, attempting to produce a clean-tasting compound that will pass safety tests. The sweetener appears to be applicable for all uses, including sweet concentrates, baked goods, and acidic soft drinks. SRI oxime V is reported to be 450 times sweeter than sucrose, without the offtaste of saccharin.[32]

Non-absorbable Leashed Polymer Sweeteners

Protein sweeteners such as monellin and thaumatin have such large molecules that they do not penetrate the taste cell, yet they are intensely sweet. This indicates that sweet taste response can be extracellular.

This information led to an innovative approach for non-absorbed sweeteners that have been under investigation since 1972 by a group of researchers at Dynapol, a high technology company in California. Their premise is that the sweetener's role has been accomplished once sweetness has been tasted. The physiological damage may come after sweeteners like cyclamates and saccharin are absorbed through the gastrointestinal tract lining and move into the blood stream and internal organs. To prevent this absorption, small molecules of the sweetening agent are leashed to much larger molecules, called polymer carriers, which are not absorbed. The sweetening agent attached to the polymer carrier passes intact through the gastrointestinal tract and is excreted.

Dynapol has been funded, partially, by the National Institute of Dental Research, to seek a non-nutritive sweetener that does not promote tooth decay. The company has also been funded by a major soft-drink manufacturer. In addition, Dynapol has been working with synthetic analogs of DHC and is attempting to develop food colors and food antioxidants, using the non-absorbable leashed-polymer technique. The company seeks to commercialize food additives "that enhance and preserve food but do not act upon internal organs." [32]

Finding a taste quality in the non-absorbed sweeteners has been a major stumbling block, and has kept Dynapol's sweetener research four years behind their progress with food colors and antioxidants. It has proven difficult to bind sweeteners to larger molecules that can reach the taste buds, have an acceptable flavor, and yet remain non-absorbable by the rest of the body. Additional problems are to make the non-absorbable sweeteners cost competitive and to clear safety tests with FDA.[33]

Non-digestible Sugar

The normal geometric arrangement of sugar is right-rotating. Under polarized light, the molecules can be seen rotating to the right. During our human evolution, we developed enzymes capable of metabolizing right-rotating sugar molecules, but not left-rotating ones. Thus, while both right- and left-rotating sugars have identical chemical components, the former is digestible and caloric, while the latter is non-digestible and non-caloric.

This difference is being explored by a company that recently acquired a patent for a sweetening process involving left-rotating sugar. However, a wide gap exists between patent acquisition and approval for marketing. Extensive studies are needed for toxicologic tests, and for physiologic effects of extra bulk from non-digestible molecules traveling through the digestive tract. To date, the left-rotating sugar has been restricted to laboratory work, which does not guarantee its economic feasibility in commercial production.[34]

Future Sweeteners: Are They Needed?

With various attempts focused on the development of alternative sweeteners, the overall problem is ignored. Resources, time, effort, and money are being diverted to futile endeavors. Ferdinand Zienty, Ph.D., from Monsanto, estimated that 8 to 10 years of research and a minimum of $10 million in funding are required to develop satisfactory safety data for any substance to be approved for the marketplace.

All of the described approaches to develop future sweeteners are flawed. Naturally derived sweeteners are not the solution, as demonstrated in the hazards of high-level consumption of licorice. Taste-altering perception of sweetness is not the answer. We lack information about possible long-range health effects from chronic distorted perception, and irreversible damage that may be inflicted on our taste receptors. There is enough evidence about artificial sweeteners, from both the recent cyclamates and saccharin fiascos and older experiences with dulcin and another early one, P4000, to make us wary of this approach. The idea of rendering undesirable substances non-absorbable is not in our best interests as consumers, although it is profitable for processors. Even if the leashing polymers pass safety tests, we should reject this approach, which encourages poor dietary patterns. There is no reason why our alimentary tracts should serve as receptacles for highly processed concoctions of low nutrient quality.

We do not need more alternative sweeteners. We do need lower consumption of *all* sweeteners and of all types of foods that require the addition of sweeteners at high levels.

The Sugar Trap:
How Can We Avoid It?

By the time children reach kindergarten age they will have seen approximately 70,000 30-second food commercials, approximately 50,000 of which are for highly sugared products. Even the best organized efforts by schools and parents are of no consequence against the powerful barrage of anti-nutritional information. It is in this context that we should interpret ABC President Frederick Pierce's recent statement that his network runs nutrition messages about 360 times a year on its children's programming. Three hundred and sixty opportunities to combat the damage wrought in one year by 10,000 sugar-drenched messages. Advertisers wouldn't bet a nickel on the chance for carrots to win over caramels.

— from the statement submitted by Action for Children's Television to the U.S. Senate, for *Food Safety: Where Are We?* U.S. Senate, July 1979

Dr. Jean Mayer, the President's advisor on nutrition and health, says if he could issue one warning to the mothers of the country it would be to cut down on all calories in everything from unnecessary desserts to soft drinks. "The human race survived very well until only a few years ago without soft drinks," declared Dr. Mayer. "What's wrong with tomato juice, orange juice, milk, and water? Nothing! And fruit is still the best dessert."

— *Washington Post*, November 2, 1969

INDUSTRIAL SMOKESTACKS, in earlier days, epitomized a prospering economy; in recent times, the symbol of pollution. Similarly, the American diet of high sugar and fat represented affluence. By the 1970s, the same diet was viewed as being related to obesity, dental decay, and directly or indirectly responsible for many other health problems.

For us in America, the 1970s was a time of greatly increased interest in food and nutrition, which resulted in new perceptions. We began to appreciate what dentists, nutritionists, researchers, and others had been trying to tell us for a long time: CUT DOWN ON SUGARS. We began to see that, from cradle to grave, sugars saturate the affluent diet.

For the first time, we questioned the soundness of adding sugar to infant feeding formulas and commercial baby foods and challenged the wisdom of using such products. We scrutinized pre-sweetened breakfast cereals intended mainly for children and questioned the morality of exposing our young, who are undiscriminating consumers in the marketplace, to the seductive sugar-laden food advertisements on television. We examined sugar use, as well as food quality, in school food programs supported with our tax money, and we questioned the appropriateness of candy- and soft drink–filled vending machines in school corridors. We became aware of the inadequacies of food labeling for sugar and other ingredients and were shocked to learn how sugars are hidden in many processed foods brought into our homes or consumed in restaurants and other feeding institutions.

Repeatedy, we were told to CUT DOWN ON SUGARS. This advice is difficult to follow for many Americans, accustomed to high sugar consumption. Many Americans learned, however, that it was possible to do so, when sugar prices soared in the mid–1970s. Home economists and food editors provided many recipes using less sweetening or no sweetening. The education committee of the Hanover [New Hampshire] Co-Op organized a "Sugar Take-Off Contest," with prizes offered for low-sugar or no-sugar recipes. The prize money was donated by local dentists.[1]

A newspaper headline in the mid-1970s read "NUTRITIONISTS DELIGHTED BY SOARING SUGAR COST." Joan Gussow, a nutrition educator, greeted the high price of sugar as

the best thing that's happened in a long time. You don't need any table sugar. We make sugar out of natural starches, and the normal person does not need any table sugar except for palatability. There is no biological need for table sugar.[2]

With the high cost of sugar by the mid-1970s, with about 80 million Americans overweight, and with heightened consumer awareness, people did begin to cut back on sugars. By 1978, a nationwide study showed that 47 percent of all American households — representing some 37 million households — had either cut back or stopped buying foods with high sugar content.[3] Food processors found a ready market for products with reduced levels of added sweeteners. Fruit canners responded to this interest by packing fruit in light, rather than in heavy syrup, and by offering some canned fruit packed in unsweetened fruit juices. The words *light* and *lite* became selling points in the marketplace.

Rarely do we eat sugars by themselves. Generally, they appear in foods and frequently are combined with starches, fats, and salt. Commonly, these four ingredients are used in many of our snacks. We have become a snacking nation, and many snack foods we choose are part of the affluent diet. Consumed at high levels, such snacks contribute to obesity and other health problems. A strong correlation has been found between increased consumption of sweetened starch-type snack foods and dental caries. Also, sucrose appears to be synergistic with salt, invoking high blood pressure.[4]

If our sugar consumption is to be reduced drastically, from cradle to grave, the program needs to be started early. The increased trend of breastfeeding is laudatory for many reasons, and among them, the practice avoids having the infant consume the sucrose or dextrose added to infant feeding formulas.

Parents discovered how easy it was to prepare solid baby foods at home, and omit the sugar. Processors understood the reason for plummeting sales of commercial baby foods and began to offer some products without added sweeteners.

It is relatively easy for parents to keep sugars away from young infants. But for growing children, as many parents have learned, the challenge is formidable. Children *can* learn desirable nutritional concepts and develop healthy food habits just as easily as undesirable ones. But in our cultural setting, the valiant attempts made to form good nutritional habits appear puny compared with the sophisticated means used to sell soft drinks, candies, pre-sweetened cereals, chewing gums, cookies, syrups, and all the rest of the sugar-laden items that are created to be particularly appealing to children.

Unfortunately, USDA aided and abetted sugar consumption by children. During 1970, the agency launched a school breakfast program that made use of "engineered" foods to replace traditional offerings of bread, fruit, and milk. Fabricated "fortified" filled cakes, "nutrified" doughnuts, peanut butter crackers, and oatmeal bars were offered. Nutritionists had grave misgivings about the soundness of this approach and charged that USDA had acquiesced to "manufacturers hungry for increased markets." Such foods did not promote the development of good food habits. "Eating a high-sugar, high-fat dessert-like product for breakfast at school is not only unhealthy, it teaches children sweet desserts are acceptable breakfast foods."[5] By 1977, USDA proposed the banning of these foods in schools but so far has not done so.

Similarly, USDA proposed prohibiting the sale of high-sugar items at schools participating in the National School Lunch Program until after the last lunch period. But this proposal, too, failed to be executed due to strong pressure from the affected food interests.[6]

In like manner, FTC's attempts to protect young children from television advertising of high-sugar foods has been scuttled, due to enormous pressure. Aware of this situation, parents and teachers need to take counteractive measures. Children can learn to enjoy non-sweet snacks and meals at school and at home. Food can provide a subject for creative learning experiences.

Sweets can be de-emphasized at birthday parties. Activities

such as corn popping or chestnut or peanut roasting can be substituted.

Items for trick-or-treat can be non-food or non-sweet. A dentist had some bookmarks printed inexpensively for distribution in a local library prior to Hallowe'en. The bookmarks had a list of suggestions for trick-or-treat offerings, including popcorn, peanuts, pumpkin seeds, sunflower seeds, soy nuts, and apples, and non-food items like balloons, erasers, pencils, and pennies.

At times, parents have resorted to dramatic gestures. Some send young children off to other homes for birthday parties wearing tags or printed T-shirts that read PLEASE FEED ME APPLES, ORANGES, OR BANANAS, BUT NO JUNK FOOD! Of course, it would be helpful to notify the party-giving parents in advance, and to offer some suggestions for suitable foods. Other parents, dismayed by well-intentioned relatives, friends, and neighbors who insist on bestowing lollipops or caramels on little Johnny as a sign of their affection, have resorted to a little white lie. "Please don't feed Johnny sweets. The doctor says he tends toward diabetes." While the sugar-pushers may disregard parents' wishes, they respect medical opinion.

For older children, various approaches are used for nutritional education. In Canada, "Snackin" signals were devised by the Canadian Automatic Merchandising Association, a group that stocks vending machines for schools as well as for industrial and commercial locations. Working with professional nutritionists and the food service industry, the association devised a Snack Coding Guide, displayed near the vending machines, with colorful "traffic signals" identification of snack foods according to their nutritional and dental acceptability. Green signifies "Go Ahead, Good Anytime," amber "Beware! Eat Only with a Meal," and red "Stop and Think! Have You Already Had a Balanced Meal?"[7]

Adults, too, need held in learning to avoid the sugar trap. Fortunately, there has been enough of a swell in consumer interest for low- or no-sugar foods for some response from processors. Some canned fruits have become available in light syrup, while others are packed in unsweetened fruit juices or water. During the mid-1970s, when sugar cost peaked, one major cola company introduced a fruit-flavored carbonated beverage that was available with or without sugar. Low-sugar fruit punches were introduced, too. More recently, some cereals, jams, and iced-tea mixes were introduced into the marketplace formulated with only half the sugar of regular products.[8]

Food co-ops have had a history of consumer education. Some have listed sugar contents of products, while others have displayed a logo of thumbs up or thumbs down, depending on the sugar levels in the products. Percentages of sugars in products would be helpful in all places where food is sold.[9] Until FDA makes this type of labeling mandatory, we need to be resourceful.

If you are a person who finds it difficult to cut down on sugar, keep a record for a week of all your food and beverage consumption. At the end of the week, check those with sugar, added either by the processor or by yourself. The next week, eliminate as many products as you can to which sugar was added by the processor. For those you personally sweetened, reduce the amount you use. If you had used two teaspoons of sugar in your coffee or tea, try using only one. Over the weeks, try to reduce the amount still further. This system of gradual reduction seems easier than "cold turkey." Ultimately, you may discover that you enjoy an unsweetened beverage more. Many individuals experience the real flavors of foods and beverages only after having reduced or eliminated added sugars. As someone remarked, "You cannot hear ordinary conversation in a boiler factory; neither can you taste natural flavors when the strong sugar taste masks them."

Once you are *aware* of your high sugar consumption, you can reduce it in various ways:

— Do you consume soft drinks? Switch to unsweetened fruit juices, tomato or vegetable juice, milk, or water.

— Do you buy canned fruit? Use products packed in their own unsweetened juices or water.

— Do you use frozen berries? Look for unsweetened products that are available.

— Do you buy prepared breakfast cereals? Choose those that have no added sugars. You can sweeten them if you want by adding fresh fruits and/or unsweetened fruit juices.

— Do you customarily eat desserts at the ends of each lunch and supper? Switch to fresh fruits, berries, and melons in season; or cheese and crackers; or nuts. Reserve baked goods and frozen desserts as occasional treats rather than as daily fare.

— Do you snack? Substitute raw vegetables such as carrots, celery, green peppers, radishes, and cauliflower for sugary or starchy snacks. Reserve a regular space for them in the refrigerator, so that they are accessible to all members of the family. If you remove soft drinks from the refrigerator, put the unsweetened fruit juices in the customary place. If you remove the candy bowl from its usual place, replace it with a bowl of fruits, nuts, or nibbles such as sunflower or pumpkin seeds.

— Do you purchase yogurt? Select plain yogurt, and add fruit if you wish, at home.

— When you are away from home, what do you buy as a quick snack? Opt for a banana instead of a candy bar.

— If your work takes you away from home, what can you eat? Brown bag your lunch and you can control its sugar content as well as other ingredients. Also, you will save money.

— If you drive long distances, what can you eat? Pack a picnic lunch before you leave home. Learn to keep some cutlery and a can opener handy in the glove compartment. Stop at any supermarket and you will find numerous items that can be eaten en route with a minimum of preparation. Purchases could include items like small wedges of cheese; individual portions of milk, cottage cheese, or yogurt; whole-grain crackers; small cans of tuna fish or sardines; individual servings of unsweetened fruit juices; bananas.

The Dietary Guidelines for Americans, issued jointly by USDA and HEW in February 1980, suggested the following, in order to avoid excessive sugars:

> use less of all sugars, including white sugar, brown sugar, raw sugar, honey, and syrups. Eat less of foods containing these sugars, such as candy, soft drinks, ice cream, cakes, and cookies. Select fresh fruits or fruits canned without sugar or light syrup rather than heavy syrup. Read food labels for clues on sugar content — if the name sucrose, glucose, maltose, dextrose, lactose, fructose, or syrups appears first, then there is a large amount of sugar. Remember, how often you eat sugar is as important as how much sugar you eat.[10]

Traps are baited, and you need to learn what the bait is in order to avoid being trapped. The sugar trap is baited with highly processed foods. Avoid them, and you avoid getting caught in the trap.

A Glossary of Sugar Terms
Notes
Selected Bibliography
Index

A Glossary of Sugar Terms

affination: Treatment of raw sugar crystals with a heavy sugar syrup to remove the adhering molasses film.

amylase: A saccharifying enzyme.

arabinose: A crystalline sugar widely distributed in plants, usually polysaccharides; found especially in gums of cherry tree, mesquite, western red cedar, sapote; produced commercially from glucose; used as a culture media for certain bacteria; also called "pectin sugar"; $C_5H_{10}O_5$.

artificial sweeteners: Synthesized, non-caloric, non-cariogenic substances, such as cyclamates and saccharin.

aspartame: L-aspartyl-L-phenylalanine methyl ester; a low-calorie sweetener synthesized from two amino acids; 150 to 200 times sweeter than sucrose; calorically equivalent to sucrose, but due to its intense sweetness, aspartame can provide the same level of sweetness as a teaspoonful of sucrose, but at only one tenth of a calorie.

Barbados molasses: A popular flavorsome molasses, containing some nutrients and usually unsulfured.

beet sugar: Sucrose; a refined carbohydrate processed from sugar beet; a disaccharide.

blackstrap molasses: Residue obtained from the manufacture of raw or refined sugar, containing some nutrients; bitter and needs to be combined with sweeter sugars; consists of more than half sucrose; a typical composition is 54 percent sucrose, 9 percent glucose, and 6 percent fructose.

blend: A solution of two or more sugars or sweeteners combined in a prescribed ratio.

blood sugar: See glucose.

bottlers' sugar: Liquid or granulated sugar that meets quality standards of the American Bottlers of Carbonated Beverages.

brix: A scale of measurement commonly used to designate the sugar concentration in a water solution, the solid's weight expressed as a percentage of the total weight.

brown sugar: A sugar with a soft grain and coated with a film of highly refined cane-flavored syrup; a by-product of the refining process; 91 to 96 percent sucrose; also called soft sugar. Liquid brown sugar was introduced in retail markets in 1978 for convenience and avoidance of caking.

buttered syrup: A type of pancake syrup containing a small amount of maple syrup blended with other sugars.

cane sugar: Sucrose; a refined carbohydrate processed from sugar cane; a disaccharide; also called sugar or table sugar.

cane syrup: An unrefined sweetener in limited production and supply made from concentrated, untreated cane, produced without use of lime, sulfur, or bleach; called pure ribbon cane syrup and sold through some health food outlets.

caramel: A food additive made from burnt sugar and used to color foods and beverages.

caramelized sugar: a brittle, brown, and somewhat bitter substance produced commercially by heating dextrose with a small amount of ammonia or ammonium salts and used as a coloring agent in carbonated beverages, bakery products, confections, and liquors; suspected as a carcinogen, possibly from the processing technique.

carbohydrates: Molecular compounds comprised of carbon, hydrogen, and oxygen.

carob: A pod from a tree in the locust family, grown mainly in the Mediterranean region. The pod, dried, toasted, and pulverized, is used as a chocolate substitute.

carob drink powder: A carob mixture to which sugar is added.

carob powder: Dried carob pod, finely ground, used as a chocolate substitute (also called St. John's-bread); used by those who wish to avoid constituents in chocolate, such as theobromine, caffeine, and fat, and also by those who need to avoid chocolate due to allergic reactions. By the mid-1970s, world supplies of chocolate were low and expensive, so many food and beverage processors turned to carob as a replacer or extender. To date, FDA has not established Standards of Identity for carob, so the quality of products varies considerably.

carob syrup: A clear colorless syrup extracted from carob, with only a faint suggestion of the chocolate flavor of toasted carob powder.

cerebrose: See galactose.

CF: See crystalline fructose; see also fructose.

char: An adsorptive and decoloring agent used in sugar refining.

comogenization: A process for producing blends with corn syrup; ingredients are combined in set ratios and subjected to heat pasteurization and fine-mesh filtration.

complex carbohydrates: Polysaccharides, found in potatoes; in grains, such as wheat and rice; and in legumes, such as peas and beans.

complex sugars: See complex carbohydrates.

confectioner's sugar: A sugar or blend pulverized to an extremely fine powder for rapid solution in cold liquids; may contain food coloring and starch to prevent caking; also called icing sugar.

cornstarch: A starchy flour made from pulverized corn; by hydrolysis, is converted to corn sweeteners such as dextrose; conventional corn syrups; and 42 percent, 55 percent, and 90 percent high-fructose corn syrups.

corn sugar: See dextrose.

corn syrup solids: Completely refined corn syrup, which is spray- or vacuum-drum-dried to lower the moisture content and form granular, semi-crystalline products; mildly sweet and moderately hydroscopic; also called glucose solids.

corn syrups: Produced commercially by hydrolysis of cornstarch and converted by the action of enzymes and/or acids into clear, concentrated aqueous solutions. Light corn syrups may contain 19 percent glucose, 2 percent fructose, and 52 percent other carbohydrates; unmixed corn syrups are only 30 percent as sweet as sucrose; enzymatically converted ones, 60 percent as sweet; also called glucose.

crystalline fructose (CF): A monosaccharide produced commercially, usually from sucrose; four calories per gram.

cyclamates: Sodium and calcium salts of cyclohexylsulfamic acid;

non-nutritive synthesized sweeteners; 30 times sweeter than sucrose; banned in the United States in 1969.

date sugar: A product made from dried, ground dates, resembling brown sugar in appearance and caking characteristic; supply is limited and erratic in health food outlets.

dextran: A sugar produced commercially by bacterial growth on a sucrose substrate; used in soft-centered confections and as a partial replacer for barley malt; also used as the principal blood-plasma expander in some hospitals; $C_6H_{10}O_5$.

dextrinize: To convert starch to dextrins.

dextrins: Substances produced commercially by hydrolysis of cornstarch and converted by the action of heat and acids or enzymes to produce maltose or glucose; used in making syrups and beer.

dextrose: Produced commercially by hydrolysis of cornstarch and converted by the action of heat and acids or enzymes to produce a sweetener used in foods and beverages, and to make caramel coloring; often blended with sucrose; about 70 to 75 percent as sweet as sucrose; also called glucose; corn sugar; grape sugar.

DHC: See dihydrochalcones.

diastase: A mixture of enzymes obtained from malt and used to convert starch to maltose; converts at least 50 times the weight of certain starches into sugars (dextrins and maltose); also called maltin.

diastatic malt: Processed malt, dried at a lower temperature than conventional malt to retain activity in its enzymes; used with baked goods, these enzymes transform starch, aid fermentation, and produce soluble proteins utilized by yeast; this malt increases nutrition, flavor, appearance, and freshness of baked goods.

dihydrochalcones (DHC): Flavones with sweetening ability — neohesperidin dihydrochalcone is 1500 to 8000 times sweeter than sucrose; prunin, about 160 times sweeter; and hesperidin and naringin dihydrochalcone, each 100 times sweeter.

disaccharide: a double sugar, consisting of two simple sugars combined and broken down during digestion to simple sugars, then absorbed; examples: sucrose, lactose, maltose.

dried corn syrup: Granulated glucose.

dried glucose syrup: Granulated glucose.

dulcin: Ethyoxyphenylurea $C_9H_{12}N_2O_2$, a non-nutritive sweetener used for more than 50 years, before long overdue tests showed it caused liver cancer in dogs; 250 times sweeter than sucrose; banned in the United States in 1950.

ethyl maltol: A form of maltol, a sweetener.

floc: An insoluble material formed in a sugar solution and that settles out slowly.

fondant sugar: A creamy mass of cooked and uncooked sugar used as base for candy and icing; dried fondant sugar is produced by direct crystallization.

formose: Synthetic sugars produced by converting formaldehyde to a complex mixture of high molecular-weight purified carbohydrates; in rat studies, animals could tolerate up to 25 percent replacement of natural sugars by formose; more was toxic possibly due to the chemical structure of these sugars, which are branched and not found in nature; not yet approved.

fructose: A very sweet natural sugar found in many fruits, honey, and as the sole sugar in bull and human semens; a monosaccharide; also called levulose; fruit sugar; produced commercially by hydrolysis of sucrose; four calories per gram; 1.2 to 2.0 times sweeter than sucrose, depending on temperature; sweetest perception is in cold, acidic beverages; in syrup form, fructose may consist of 72 percent fructose and 6 percent glucose; in crystalline form, 96 to 100 percent fructose, with traces of glucose; $C_6H_{12}O_6$.

fructose syrup: A misleading term for high-fructose corn syrup, derived from corn.

fruit sugar: See fructose.

galactose: A sugar produced commercially in right-rotating form by hydrolysis of lactose melibiose, raffinose, or certain polysaccharides such as agar and pectin; produced in left-rotating form from flaxseed mucilage; less soluble and less sweet than glucose; about 58 percent as sweet as sucrose; used medically in liver function test; also known as brain sugar; cerebrose; $C_6H_{17}O_8$.

glucose: A sugar that occurs naturally and in a free state in fruits and other parts of plants, in polysaccharides, cellulose, and starch; found in glycogen and in human blood; a main source of energy for living organisms; produced commercially by hydrolysis of cornstarch; a monosaccharide; also called dextrose; blood sugar; grape sugar; corn sugar; only 20 percent as sweet as sucrose; $C_6H_{12}O_6$.

glucose solids: A nearly colorless substance made from starch or a starch-containing substance; when made from corn, known as corn syrup solids.

glucose syrup: A liquid solution of glucose, produced commercially by hydrolysis of cornstarch.

glycine: A naturally occurring amino acid that formerly was used to mask saccharin's bitterness; had been on the GRAS list, but was later removed after experiments demonstrated adverse effects: At high levels, glycine given to rats suppressed growth, led to body weight loss, increased liver weight, and in some instances, caused cervix paralysis; further glycine use with saccharin was forbidden; $C_2H_5NO_2$.

granulated sugar: Refined sucrose in crystallized form.

grape sugar: Glucose.

hexahydroxyl alcohol: See sugar polyols.

hexitols: See sugar polyols.

HFCS: See high-fructose corn syrups.

high-fructose corn syrups (HFCS): Produced commercially by hydrolysis of dextrose and converted by the action of heat and enzymes to fructose, and then further processed to produce high-fructose corn syrups; sweeter than invert sugar, which it frequently replaces; four calories per gram; 42 percent HFCS is as sweet as sucrose; 55 percent HFCS, 1.1 to 1.15 times sweeter; 90 percent HFCS, 1.6 times sweeter.

honey: A natural sweetener produced from gathered nectars and converted by the enzymes of the bee into invert sugar; compositions and flavors vary depending on the nectar sources; among its main components are two sugars, fructose and glucose, with smaller amounts of sucrose and maltose; honey is 1.3 times sweeter than sucrose; one tablespoon of honey contains 65 calories; sucrose, 50 calories.

hydrolysis: A chemical reaction in water, in which a bond in the reactant other than the water is split and hydrogen and hydroxyl are added, with two or more new compounds formed.

hydrometer: A calibrated floating instrument used to determine a liquid sweetener's density.

hydroscopic: Absorbing and retaining moisture.

icing sugar: See confectioner's sugar.

imitation maple syrup: Also called pancake blend or syrup; waffle blend or syrup.

inulin: A tasteless polysaccharide that occurs instead of starch in many plants, especially in the tubers or roots of Jerusalem artichoke, dahlia, and chicory; produced commercially by hydrolysis to yield fructose; used as a fructose source and medically in a kidney function test.

inversion: A specific type of hydrolysis that produces invert sugar.

invertase: An enzyme capable of converting sucrose to invert sugar.

invert sugar: Digested sugar; a mixture of 50 percent each of sucrose and dextrose, produced commercially by hydrolysis; only slightly hydroscopic; 1.0 to 1.3 times sweeter than sucrose; used mainly as a syrup in foods, and medically in intravenous feeding solutions.

invert sugar syrup: A liquid solution of invert sugar.

lactate esters: Sugar-derived solvents that are highly miscible; considered as a naturally occurring constituent of beverages and foods. Traces are identified in California and Spanish sherries. One form, ethyl lactate, is used as a flavoring agent.

lactose: Milk sugar, occurring in the milk of all mammals; a disaccharide; produced commercially from whey and skim milk and used by food and beverage processors and the pharmaceutical industry. Lactose is from 15 to 32 percent as sweet as sucrose. Human milk consists of 7 percent lactose; cow's milk, 4.8 percent.

lactulose: An isomerized form of whey lactose, more soluble and sweeter than lactose; 48 to 62 percent as sweet as sucrose but is non-nutritive. Its humectant property may make it useful as a sucrose substitute in some cases, but due to its laxative property, lactulose cannot be used at high levels.

levulose: One of the two simple sugars formed in sucrose inversion; a monosaccharide; also called fructose; fruit sugar; about 1.3 to 1.8 times sweeter than sucrose.

licorice: Glycyrrhiza glabra; the long, thick, sweet roots of a perennial leguminous plant from the Mediterranean region; the source of licorice extract, used to flavor foods, beverages, and tobacco and to mask unpleasant flavors in drugs; ammoniated glycyrrhizin is about 50 times sweeter than sucrose; one pound of licorice added to 100 pounds of sucrose yields a sweetness level of 200 pounds of sucrose.

liquid sugar: Sugar in solution; sucrose in water; the common form of sugar used by food and beverage processors because liquid can be blended easily with other ingredients, and bulk handling is economical.

liquor: A term applied generally to partially concentrated sugar solutions and syrups.

malt: A sweetener produced commercially by steeping grain (usually barley) in water to soften and germinate; enzymes convert the grain's starch into sugar; the resulting form is kiln-dried, ground, and used as a nutrient, and by brewers and distillers.

malt extract: A powder processed from barley, containing diastase, dextrin, dextrose, protein, and salts; used for baked goods, processed cereals, and confections; currently under safety review.

malt sugar: Maltose.

malt syrup: A non-diastatic syrup extracted from barley malt and concentrated into a liquid; currently under safety review.

maltin: See diastase.

maltodextrins: Hydrolyzed carbohydrates; production similar to that of corn syrups, but the conversion is arrested at an earlier stage.

maltol: A crystalline compound found especially in the bark of young larch trees, in pine needles, chicory, wood tars and their oils, and in roasted malt; used as a flavoring agent to impart a "fresh baked" odor and flavor to bread and cake; $C_6H_6O_3$.

maltose: Malt sugar produced commercially by hydrolysis of starch;

a disaccharide, fermentable, reducing sugar used in brewing, distilling, and in processed foods; also called malt sugar; only 20 to 30 percent as sweet as sucrose; $C_{12}H_{22}O_{11}H_2O$.

maltose syrup: A liquid solution of maltose.

manna sugar: See mannitol.

mannitol: A rare sugar found as mannose in chicory and in roasted malt; also called sugar polyol; sugar alcohol; about as sweet as dextrose; about 50 percent as sweet as sucrose; $C_6H_{14}O_6$.

mannose: A crystalline sugar, the principal constituents of manna ash, which is reduced to mannitol; $C_6H_{12}O_6$.

maple sugar: A granulated form of maple syrup, made by boiling the syrup to the hard sugar stage and immediately stirring to promote crystallization.

maple syrup: Concentrated sap from the maple tree, mainly sucrose, with water and small amounts of invert sugar and malic acid; produced by evaporating the surplus water; 40 calories per tablespoonful.

melibiose: A disaccharide sugar formed by partially hydrolyzing raffinose, by fermentation with top yeast, which removes the fructose; also called molasses sugar; $C_{12}H_{22}O_{11}$.

milk sugar: Lactose; a disaccharide; 30 percent as sweet as sucrose.

miraculin: An intensely sweet glycoprotein in the miracle berry, *Synsepalum dulcificum,* with a large molecule and a molecular weight of 42,000; the sweetness of miraculin is 2500 greater than an equivalent weight of sucrose.

molasses: Syrup obtained by evaporation and partial inversion of clarified or unclarified sugar cane juice.

monosaccharide: A sugar of simple molecular structure; examples: arabinose, dextrose, galactose, glucose, levulose, mannose, xylose.

muscovados: Unrefined raw sugar obtained from sugar cane juice.

Nigerian berry: Dioscoreophyllum cumminsii; a plant containing a principle that is 2500 times sweeter than sucrose; also called serendipity berry; molecular weight is about 44,000.

non-diastatic malt: A syrup extracted from barley malt and concentrated into syrup; also known as malt syrup.

non-nutritive sweeteners: Synthesized non-caloric sweeteners, such as cyclamates and saccharin.

oligosaccharides: Any of the saccharides that contain a known small number of monosaccharide units and include especially the disaccharides, trisaccharides, and tetrasaccharides; example: glucose syrup.

pancake blend: A mixture of syrups that may or may not contain a small percentage of maple syrup; imitation maple syrup; waffle blend or syrup; pancake syrup.

pancake syrup: Pancake blend.

pectin sugar: A product obtained as a powder or syrup, extracted from citrus peel, dried apple powder, or dried sugar beet slices; used chiefly in jellies, pharmaceuticals, and cosmetics; also called arabinose.

pineapple syrup: Clarified, concentrated fruit sugar derived from by-products of pineapple processing, namely the shells and outer portions of the pineapple; this product may be declared as "pineapple syrup" on label ingredients of food products.

polysaccharides: Multiple sugars consisting of simple sugars that are combined during plant development and growth; found in grains, dried peas, beans, rice, legumes, potatoes, unripe bananas and apples, old sweet corn, glycogen, cellulose, and hemicellulose.

potato starch sugar: Produced commercially by hydrolysis using enzymes with a by-product of potato processing. Potato starch slurry, called "white water," from potato cuttings is converted to a syrup for use on potatoes before they are fried.

raffinose: A crystalline, slightly sweet, non-reducing trisaccharide sugar found in sugar beet, eucalyptus, manna, and many cereals; produced commercially by water extraction from cottonseed meal, or by hydrolysis, yielding right-rotating forms of fructose, galactose, and glucose; invertase splits raffinose into melibiose and saccharose; raffinose, present in the soybean, produces flatulence in the human intestinal tract; $C_{18}H_{32}O_{16}$.

rare sugars: A group of sugar alcohols, such as mannitol, sorbitol, and xylitol; also called sugar polyols; hexahydroxyl alcohols; hexitols.

raw sugar: Newly formed crystals, coated with molasses, that result from the boiling of sugar cane juice; also called muscovados.

reducing sugars: A general term for certain sugars, such as dextrose, levulose, and others that are easily oxidized by alkaline copper sulfate.

refined sugar: See sucrose and dextrose.

refiners' syrup: The residual liquid product obtained in the processing of refined sugar and used by food processors but described as having "such a salty taste and such a peculiar flavor as to be practically inedible."

remelts: A term applied to sugars obtained by reboiling sugar liquors.

rice syrups: Produced from rice, barley malt, and water. Enzymes in the barley malt convert the rice starch into complex sugars without use of acid. Such products are sold in some health food outlets.

saccharometer: A hydrometer calibrated in percent solids to determine the solids in a sugar solution.

saccharides: Carbohydrates consisting of molecular compounds of carbon, hydrogen, and oxygen.

saccharify: To hydrolyze into a simple soluble fermentable sugar by means of the enzyme amylase.

saccharin: 2,3-Dihydro-3-oxobenziso sultonazole; currently the only non-nutritive synthesized sweetener permitted but whose ultimate fate is uncertain; 300 to 500 times sweeter than sucrose; $C_7H_5NO_3S$.

saccharose: Any of the compound sugars including disaccharides and trisaccharides.

serendipity berry: See Nigerian berry.

simple sugars: Monosaccharides; sugars that dissolve readily, are quickly available for absorption from the digestive tract into the bloodstream, and supply energy rapidly; examples include honey, maple sugar, and fruit sugars. Two simple sugars combined produce a disaccharide, which requires slightly more breakdown time prior to absorption; examples include lactose and sucrose. Large quantities of simple sugars raise the blood sugar (glucose) too rapidly, stress the pancreas, which in turn may overreact and excrete excessive insulin, causing rapid blood sugar reduction that may result in many unpleasant symptoms, including headache, fatigue, dizziness, weakness, and depression.

soft sugar: Sugar having a soft grain and coated with a film of highly refined cane-flavored syrup; brown sugar; a by-product of the refining process.

sorbitol: A rare sugar, first found as sorbose in the ripe berries of the mountain ash but also occurring in many other berries, fruits, seaweeds, algae, and blackstrap molasses; produced commercially from glucose or corn sugar; only 50 to 70 percent as sweet as sucrose but with the same number of calories; $C_6H_{14}O_6$.

sorbose: A crystalline sugar reduced to sorbitol by fermentation and used mainly for making ascorbic acid; $C_6H_{12}O_6$.

sorghum: A syrup made from sorghum grain that resembles sugar cane but contains a high proportion of invert sugars, starch, and dextrin.

stachyose: A sweet crystalline sugar found especially in the tubers of Chinese artichoke, and by hydrolysis yields glucose, fructose, and galactose; a tetrasaccharide. This sugar, present in the soybean, produces flatulence in the human intestinal tract.

sucrose: Cane or beet sugar; a disaccharide composed of two simple sugars, glucose and fructose; also called table sugar; refined sugar; white sugar; four calories per gram; 99.9 percent pure. Sucrose is the standard used for measurement of sweetness levels of all sugars; $C_{12}H_{22}O_{11}$.

sucrose palmitate: Sugar converted from starch by cereal malt enzymes, used as a sugar replacer in baked goods since it supports the production of carbon dioxide for leavening more efficiently than sucrose. Baked goods may be labeled "sugar free" when sucrose palmitate is used in the baking formula.

sugar alcohol: See sugar polyols.

sugar beet extract flavor base: Concentrated residue of soluble sugar beet extractives from which sugar has been recovered; approved for use as a food additive but not listed on food labels.

sugar polyols: Hexahydroxyl alcohol, poorly absorbed in the body; examples are mannitol, sorbitol, xylitol.

sugar solids: The total sugar present in a solution. Commercial liquid sugars and blends sometimes are purchased on a dry sugar solids basis.

sugar syrup: A liquid solution of sucrose or other sweeteners.

synthesized sweeteners: See artificial sweeteners.

table molasses: The liquid component resulting from sugar refining.

table sugar: Sucrose, from cane or beet sugar.

terpenoids: Substances that are from 50 to 2000 times sweeter than sucrose; not yet approved.

tetrasaccharides: Any of a class of carbohydrates that, on complete hydrolysis, yield four monosaccharide molecules; example: stachyose.

thaumatin I and II: Proteins isolated in 1972 from a plant, *Thaumatococcus daniellii,* known as katemfi or "miraculous fruit of the Sudan." The large-molecule proteins are 1600 times sweeter than sucrose; molecular weight, about 21,000.

total invert sugar: A mixture of glucose and fructose, formed by splitting sucrose in a process called inversion, accomplished with acids or enzymes; sold in liquid form to food processors; sweeter than sucrose; helps keep baked goods and confections fresh; and prevents foods from shrinking.

treacle: Produced from granulated sugar liquors.

trisaccharide: Any of a class of sugars that, on complete hydrolysis, yields three monosaccharide molecules; example: raffinose.

turbinado sugar: Partially refined cane sugar that has been washed, dried, but not yet bleached; it is an off-white yellow or gray; used by food processors. 99 percent sucrose; edible, if properly processed; four calories per gram. The word *turbinado* comes from the turbine action of the centrifuge as the sugar is sprayed with water. In the past, some samples contained contaminants.

waffle blend: See imitation maple syrup.

waffle syrup: See imitation maple syrup.

washed raw sugar: Sugar after treatment by affination. See affination.

water-white: A sugar solution that is clear and colorless.

white sugar: See sucrose.

wood sugar: Xylose.

xylan: A polysaccharide built from right-rotating xylose units and occurring in association with cellulose.

xylitol: A sugar polyol found as xylose in numerous food plants; about as sweet as sucrose, with the same number of calories, namely four calories per gram; $C_5H_{10}O_5$.

xylose: A crystalline sugar widely distributed in plant material, especially in wood, straw, and hulls; not found in a free state but as xylan; $C_5H_{10}O_5$.

Notes

Note references that appear out of sequence in any chapter refer to previously cited notes within the same chapter.

Chapter 1. The Sugar Trap: How Did We Get into It?

1. *New York Times,* March 4, 1980. Little is known about the biophysical and chemical mechanisms involved in sweetness. Why are such diverse molecules as saccharin, maltol, glycine, chloroform, lead acetate, and some beryllium salts sweet? Only a few single sugars (monosaccharides) and multiple sugars (polysaccharides) are sweet. Why is cotton, a polysaccharide, tasteless, while corn, another polysaccharide, is sweet? Nor is the basis of our sweetness perception understood. How does the sweet molecule act positively with our tongue's taste receptors? Our tongues have more than a half million taste cells, clustered in groups of about 50 cells to form taste buds. The ends of the buds are composed of tiny fingerlike projections, called microvilli. Tastes occur at the microvilli that are constantly bathed in saliva. Strangely, taste can be perceived with some substances by intravenous stimulation, not in direct villi contact. Formerly, it was thought that specific taste nerves existed capable of responding to only one type of stimulus, thus giving rise to the idea of primary tastes of sweet, bitter, salt, and sour. Now it is believed that although differences exist among taste receptor cells, most are not specialized but respond to several tastes. Individuals differ in taste perceptions, and genetics may be a factor. Some individuals perceive sweetness intensity of sugars at lower levels than others, while some are more sensitive than others to saccharin's bitterness.

2. *Science,* February 27, 1976.

3. *New York Times,* June 5, 1979.
4. In a survey at London Hospital Medical College, 77 percent of all meals and snacks fed to infants eight to eleven months old contained sucrose (*New Hampshire Nutrition Alert,* Bread and Task Force, December 1978). The Department of National Health and Welfare, Canada, analyzed 24 infant cereals and found that sucrose levels, although lower than those in adult breakfast cereals, were up to 21 percent; the combined level of all sugars, up to 27 percent. Only 10 of the infant cereals contained less than 10 percent sucrose (*Journal of the Canadian Dietetic Association,* January 1979).
5. *Science,* November 14, 1975.
6. *World Review of Nutrition and Dietetics,* Vol. 22, 1975.
7. *The Professional Nutritionist,* Winter 1981.
8. *Historical Statistics of the United States, Colonial Times to 1959,* U.S. Bureau of Census, 1960. In 1874, processed sugar consumption in the United States was at 40 pounds. According to historical studies, 200 years ago, people consumed only four pounds of sugar from all sources yearly ("Working Papers," *Commonweal Research Publication,* Vol. 1, No. 3, Bolinas, California). Four pounds yearly is quoted also in *Nutrition Reviews,* Vol. 32, 1974; and John Yudkin, M.D., *Sweet and Dangerous,* Wyden, 1972, estimated the annual U.K. consumption of sugar 200 years ago to be about four to five pounds.
9. *Agricultural Research,* June 1973.
10. Louise Page and Berta Friend, "Level of Use of Sugars in the United States," in *Sugars in Nutrition,* Horace L. Sipple and Karen W. McNutt, eds., Academic Press, 1974.
11. John L. and Karen Hess, *The Taste of America,* Grossman, 1977; *Sweeteners Used by the Baking Industry,* Agricultural Economic Report 32, USDA, May 1963.
12. Berta Friend, *Changes in Nutrients in the United States Diet Caused by Alterations in Food Intake Patterns,* ARS, USDA, 1974.
13. *Food Consumption Prices and Expenditures,* Agricultural Economics Report 138, USDA, September 1979.
14. *Nutrition News,* July 1975.
15. Sucrose exerts synergistic effects with such other ingredients

as refined flours, and salt, commonly found at high levels in many processed foods. A strong correlation exists between the increased consumption of sweetened starch-type snack foods and dental caries and has occurred even when the per capita sugar consumption level has remained constant (Dr. Abraham Nizel, *Edible TV*, Select Committee on Nutrition and Human Needs, U.S. Senate, September 1977). When sucrose and salt were combined in the diet of experimental animals, sucrose had a synergistic effect and produced more hypertension than did salt alone (*American Journal of Clinical Nutrition*, March 1980).

16. *Consumer Affairs Newsletter*, City of Syracuse, N.Y., May 1979.
17. *New York Times*, December 20, 1978.
18. *Food Engineering*, March 1981.
19. Nick Mottern, *Guidelines for Food Purchasing in the United States*, U.S. Senate, Select Committee on Nutrition and Human Needs, 1978.
20. *Why Sugar?* The Sugar Association, Inc., undated; *Food Processing*, July 1978, September 1977; *Food and Drug Packaging*, February 22, 1979; *Restaurant Business*, August 1, 1980.
21. *American Journal of Clinical Nutrition*, April 1974; *Nutrition Today*, Spring 1969; *Nutrition Abstracts and Reviews*, May 1977, September 1970; *American Heart Journal*, May 1975; *Medical World News*, February 12, 1971; *Drug Therapy*, October 1976.
22. *Nutrition Reviews*, September 1968; Marc LaLonde, *A New Perspective on the Health of Canadians*, Minister of National Health and Welfare, Canada, 1974.

Chapter 2. Refined Sweeteners: Sweet and Detrimental? Cane, beet, and corn sugars

1. Allan G. Cameron, *Food Facts and Fallacies*, Faber and Faber, 1971; Reay Tannahill, *Food in History*, Stein and Day, 1973; Peter Farb and George Armelagos, *Consuming Passions, the Anthropology of Eating*, Houghton Mifflin, 1980.

2. James Rorty and N. Philip Norman, M.D., *Food for Tomorrow*, Devin Adair, 1956.

3. *World Review of Nutrition and Dietetics*, Vol. 22, 1975.

4. W. W. Duke, *Asthma, Hay Fever, Urticaria, and Allied Manifestations of Allergy*, C. V. Mosby, 1926.

5. *Journal of Experimental Medicine*, Vol. 70, 1939; *American Journal of Hygiene*, Vol. 34, 1941; *Journal of Laboratory and Clinical Medicine*, Vol. 36, 1950.

6. Theron G. Randolph, M.D., "The Role of Specific Sugars," in *Clinical Ecology*, Lawrence D. Dickey, M.D., ed., Charles C. Thomas, 1976.

7. *Journal of Laboratory and Clinical Medicine*, Vol. 36, 1950, cited previously.

8. *Annals of Allergy*, Vol. 9, 1951.

9. *Evaluation of the Health Aspects of Sucrose as a Food Ingredient*, FASEB report to FDA, U.S. Department of Commerce, National Technical Information Service, 1976.

10. *Health and Human Services News*, December 30, 1980.

11. *American Journal of Clinical Nutrition*, January 1978.

12. Reiser and Szepesi's discussion of Yemenite Jews was based on the testimony of Aharon M. Cohen, M.D., before the Hearings of the Select Committee on Nutrition and Human Needs, U.S. Senate, April 30, 1973.

13. O.B. Wurzburg, "Starch in the Food Industry," in *Handbook of Food Additives*, Thomas E. Furia, ed., Chemical Rubber Co., 1968.

14. Harvey W. Wiley, M.D., *An Autobiography*, Bobbs Merrill, 1930. In short-term feeding with glucose syrup, one gross abnormality noted in rats was cecums enlarged to approximately twice their normal size (*British Journal of Nutrition*, Vol. 29, 1973).

15. Harvey W. Wiley, M.D., *The History of a Crime Against the Food Law*, self-published, 1929.

16. Oscar E. Anderson, Jr., *The Health of a Nation*, University of Chicago Press, 1958.

17. *Journal of the Kansas Medical Society*, Vol. 27, 1936; *Annals of Allergy*, Vol. 7, 1949; H.J. Rinkel, M.D., instructional course, American College of Allergists, 1944.

18. *Annals of Allergy,* Vol. 7, 1949, cited above.
19. G. G. Birch, L. F. Green and C. G. Coulson, eds., *Glucose Syrups and Related Carbohydrates,* Elsevier, 1970; *American Journal of Diseases of Childhood,* Vol. 49, 1935.
20. *Archives of Surgery,* Vol. 61, 1950.
21. *Sugar and Sweetener Report,* USDA, December 1967; December 1980.
22. *Chemical and Engineering News,* April 11, 1977.

Chapter 3. Traditional Sweeteners: Overrated? Raw and brown sugars, molasses, honey, maple syrup, sorghum, malt, grain syrups, whey

1. *Food Facts from Rutgers,* July/September 1973.
2. *Journal of the American Medical Association,* July 10, 1972.
3. *Restaurant Hospitality,* October 1978, November 1978; *Institutions,* July 15, 1980. Retail prices for raw sugar are triple those of table sugar. In a 1974 price comparison, table sugar was selling for 16 cents a pound; raw sugar, in the natural food section of the same store, for 48 cents a pound (Dr. Jean Mayer's column, *Concord* [N.H.] *Monitor,* November 1, 1974). In a flier distributed to a group called Organic Merchants, Inc., health food store personnel were informed that "No Organic Merchant sells white sugar or any product containing white sugar because it is a foodless food ... Organic Merchants do not sell brown sugar or 'raw' sugar or any products containing brown sugar either, because the plain fact is that brown sugar is a shuck (for those not familiar with the term, let's call brown sugar phony). It does not seem to me to be good judgment to ban white sugar because it is refined ... and then turn around and sell a product which is made from ... the very same white sugar" (Fred Rohe, *The Sugar Story,* distributed by Organic Merchants, Inc., undated).
4. John Yudkin, M.D., *Sweet and Dangerous,* Wyden, 1972.
5. *New York Times,* December 6, 1974.
6. Dr. Jean Mayer's column, *Concord* [N.H.] *Monitor,* June 19, 1980, February 11, 1977.

7. *Consumer Reports,* September 1972.

8. *Food Processing,* December 1978; *Food Product Development,* October 1978.

9. *Cancer Research,* May 1965.

10. *Baking Industry,* May 1976.

11. *Food Processing,* May 1975, November 1975; *Baking Industry,* July 1976, August 1977; *Baking Production and Management,* November 1972; *Snack Food,* April 1976.

12. *Health, Education, and Welfare News,* September 21, 1978.

13. *All About Molasses,* American Molasses Co., 1952.

14. *Food Processing,* January 1979.

15. *Processed Prepared Foods,* October 1979; *Food Processing,* May 1980; *Snack Food,* October 1979.

16. The composition of one tablespoon of average light molasses and blackstrap molasses is, respectively: food energy — 50.0 calories, 43.0; carbohydrates — 13.0 grams, 11.0; calcium — 33.0 milligrams (mg), 137.0; phosphorus — 9.0 mg, 17.0; iron — 0.9 mg, 3.2; sodium — 3.0 mg, 19.0; potassium — 183.0 mg, 585.0; no vitamin A; thiamin — 0.01 mg, 0.02; riboflavin — 0.01 mg, 0.04; niacin — trace, 0.04 mg; ascorbic acid — none (*Nutritive Value of American Foods in Common Units,* Agricultural Handbook No. 456, ARS, USDA, November 1965). Molasses samples analyzed by atomic absorption spectrograph for parts per million (ppm) of trace elements showed the following nutrients in cane molasses and beet molasses, respectively: zinc — 1.08 ppm, 3.02; copper — 0.55, 1.75; iron — 20.4, 30.2; manganese — 4.14, 5.83; chromium — 0.08, 0.30. Although beet molasses was higher than cane in all trace elements measured, beet molasses is limited for use in cattle feed (Carl C. Pfeiffer, Ph.D., M.D., *Mental and Elemental Nutrients,* Keats Publishing, 1975). In another molasses analysis, the following elements were found in ppm: magnesium, 250; chromium, 1.21; manganese, 4.24; cobalt, 0.25; copper, 6.83; zinc, 8.3; and molybdenum, 0.19 (Henry A. Schroeder, M.D., *The Trace Elements and Man,* Devin Adair, 1973).

Hydroxymethyl furfural (HMF), a highly reactive and allergenic group of compounds, are major decomposition products of sugar, with some formed by cooking foods that contain sugars. Gas chromatography shows that those formed in acid solutions usually are derived from furane or pyrane and include a wide variety of somewhat complex chemicals. Those formed in alkaline media are even more complex, including carbon ring compounds with a variety of functional groups attached. The conditions for these compounds to form are especially favorable in processing sugar cane juice to form blackstrap molasses. Chemists view its coloring matter as water-soluble tar. Dr. Alsoph Corwin, chemist emeritus from Johns Hopkins, noted that "Blackstrap molasses is widely recommended as a 'health food' because it contains residual minerals which are removed from cane juice in the purification of cane sugar. Due consideration for food stresses, however, would dictate that the necessary minerals be obtained from a less risky source. So-called 'raw sugar,' turbinado sugar, and brown sugar all contain varying amounts of molasses and all should be considered as rich sources of chemical contaminants formed by the cooking process. While some of the colored materials present in cooked sugar solutions have been identified, many of them are so complex that their structures are still unknown. All of them are artificial chemical contaminants" (Dr. Alsoph Corwin, *The Most Common Source of Chemical Contaminants,* undated paper).

17. Invert sugar, processed for food and beverage use, is made from sucrose treated with acid or enzymes to split the disaccharide into its separate components. This process, called inversion, results in the commercial "invert sugar" on food labels. Its special properties of interest to processors are its resistance to crystallization and its moisture-holding capacity; both are useful in certain food formulations.

18. *American Agriculturist,* April 1965, December 1965.

19. *American Bee Journal,* Vol. 95, 1955.

20. *Beekeeping Bulletin* No. 5, Department of Agriculture, Florida, January 1958.

21. Anstice Carroll and Embree de Persiis Vona, *The Health Food Dictionary,* Prentice-Hall, 1973.

22. *Whole Foods' Natural Foods Guide,* And/Or Press, 1979.

23. *The Washingtonian,* Winter 1978/1979.

24. Frederick Accum, *Death in the Pot,* London, publisher unlisted, circa 1830.
25. *FDA Enforcement Report,* December 8, 1976, March 2, 1977, May 21, 1980, October 22, 1980; *FDA Consumer,* February 1979, November 1979, February 1980.
26. *Publications and Patents,* Eastern Regional Research Center, ARS, USDA, July/August 1979; *Chemical and Engineering News,* January 30, 1978; *Journal of the Association of Official Analytical Chemists,* Vol. 62, 1979; *Annals of Chemistry,* February 1979; *American Bee Journal,* July 1977.
27. *Whole Foods,* December 1979.
28. *Processed Prepared Foods,* September 1979, January 1980; *Snack Food,* October 1979; *Food Product Development,* May 1980; *Food Processing,* May 1974, July 1975; *Food Engineering,* February 1974, January 1980; *Bakers' Digest,* August 1974.
29. *Health Food Retailing,* February 1980, April 1980.
30. *Food Processing,* September 1974, August 1975, May 1976; *Baking Industry,* October 1976, May 1977; *Food Technology,* August 1974; *Food Engineering,* March 1976. An imitation honey was patented in Japan, consisting of a 30 percent glucose solution, treated with enzymes, pollen, sodium citrate, royal jelly, and six organic acids — gluconic, citric, malic, succinic, lactic, and formic (*Modern Nutrition,* May 1964).
31. The pesticides most lethal to bees are Sevin, DDT, Lindane, Chlordane, DNC (4,4'-Dinitrocarbanilide), and Dieldrin. DNOC (4,6-Dinitro-*o*-cresol) is less toxic. Honey is unlikely to be contaminated with the insecticides most toxic to bees because they die before their honey is stored in the hive. One exception is Schradan (octamethyl pyrophosphoramide), which is not toxic to bees. This systemic insecticide is applied to plants before they flower, appears in the nectar, and may be carried back to the hive. Schradan does not decompose and may be stored in honey. Up to 44.5 ppm of Schradan was found in honey samples, which is equivalent to about 20 mg of this pesticide in a pound of honey (Franklin Bicknell, M.D., *Chemicals in Food and in Farm Produce: Their Harmful Effects,* Faber and Faber, 1960).

208   NOTES

208   NOTES

32. *Toxicants Occurring Naturally in Foods,* Publication 1354, NAS/NRC, 1966; USDA press release, October 15, 1970; Bicknell, *Chemicals in Food and in Farm Produce,* cited above.
33. *Science,* February 4, 1977.
34. *Bulletin of the Faculty of Medicine,* Istanbul, Vol. 12, 1949; *Archives of Biochemistry and Biophysics,* Vol. 79, 1959; Bicknell, *Chemicals in Food and in Farm Produce,* cited above.
35. *Infectious Diseases,* September 1979; *Science News,* July 15, 1978; *New York Times,* June 25, 1978, July 7, 1978; *Medical Tribune,* September 20, 1978; *Science,* September 1, 1978.
36. *California Morbidity,* July 7, 1978, July 14, 1978; *Morbidity and Mortality Monthly Report,* Center for Disease Control, USPHS, January 20, 1978, October 20, 1978.
37. Fergus M. Clydesdale and Frederick J. Francis, *Food, Nutrition and You,* Prentice-Hall, 1977.
38. The following nutrient values are listed for one tablespoon of average honey: food energy, 64.0 calories; 17.3 grams carbohydrates; 1.0 mg calcium; 1.0 mg phosphorus; 0.1 mg iron; 1.0 mg sodium; 11.0 mg potassium; no vitamin A; trace of thiamin; 0.01 mg riboflavin; 0.1 mg niacin; and a trace of ascorbic acid (*Nutritive Value of American Foods in Common Units,* previously cited). Honey samples, analyzed by atomic absorption spectrograph for ppm of trace elements, showed the following nutrients in orange blossom honey and clover honey, respectively: zinc — 0.98 ppm, 0.82; copper — 0.05, 0.08; iron — 0.59, 1.81; manganese — 0.17, 0.29; and chromium — 0.02, 0.05 (Pfeiffer, *Mental and Elemental Nutrients,* previously cited).
39. Helen and Scott Nearing, *The Maple Sugar Book,* John Day, 1950; *American Agriculturist,* January 1979.
40. *That We May Eat,* USDA Yearbook, 1975.
41. *American Agriculturist,* October 1965; C. O. Willitz, *Maple Syrup Producers Manual,* Agriculture Handbook 134, USDA, rev. 1965.
42. *Consumer Bulletin,* September 1962.
43. *American Agriculturist,* February 1966.
44. *American Agriculturist,* March 1978; *Consumer Reports,* May 1979.

45. *Let's Live,* June 1975.

46. *Chemical and Engineering News,* May 22, 1972.

47. *New York Times,* June 30, 1975; *Consumer Reports,* October 1967, October 1968.

48. *New York Times,* May 4, 1975.

49. *Consumer Reports,* May 1979.

50. *Vermont Freeman,* August 1973.

51. *Food Product Development,* July 1979.

52. *Media and Consumer,* June 1975.

53. *FDA Enforcement Report,* February 27, 1980, May 21, 1980, July 16, 1980; *FDA Consumer,* February 1979, June 1980, September 1980.

54. *Institutions,* March 15, 1979; *Food Processing,* February 1976, September 1977, January 1979; *American Dairy Review,* July 1976; *Plant Food Ideas,* June 1977, December 1977; *Baking Industry,* May 1977; *Dairy and Ice Cream Field,* September 1979.

55. *Food Processing,* February 1978.

56. *Consumer Reports,* May 1979.

57. *Brattleboro* [Vt.] *Daily Reformer,* December 6, 1975.

58. *Agricultural Research,* USDA, March 1972.

59. *Service USDA,* September 1976.

60. *Christian Science Monitor,* November 19, 1973.

61. *Snack Food,* July 1978.

62. *Snack Food,* October 1979.

63. *Food Product Development,* October 1976, January 1980; *Food Processing,* September 1978; *Food Engineering,* April 1980, September 1980.

64. *Food Engineering,* September 1979, September 1980; *Food Product Development,* September 1980.

65. *Food Product Development,* November 1980.

66. *Food Product Development,* May 1978, January 1980; *Dairy Field,* June 1980; *Dairy Record,* October 1979; *Food Engineering,* July 1980.

Chapter 4. New Sweeteners: An Unjustified Craze? Crystalline fructose, high-fructose corn syrups, aspartame, polydextrose

1. T. E. Doty and E. Vanninen, *Fructose, A Review of Nutritional, Medical, and Metabolic Aspects,* Finnish Sugar Co., undated.
2. *Food Technology,* November 1975.
3. *Processed Prepared Foods,* February 1979.
4. *Dietary Sugars in Health and Disease: Fructose,* Federation of American Societies for Experimental Biology, report to FDA, October 1976.
5. *Family Circle,* February 20, 1979.
6. *Snack Food,* February 1980.
7. *Washington Post,* October 11, 1979.
8. *Alive,* Autumn 1979.
9. *Food Engineering,* March 1979.
10. *Health Food Business,* August 1979.
11. *Whole Foods,* May 1979.
12. *Health Food Retailing,* July 1979.
13. *New York Times,* May 7, 1980.
14. *New York Times,* January 4, 1981.
15. *Food Engineering,* October 1980.
16. *Diabetes Care,* July/August 1978.
17. *American Journal of Clinical Nutrition,* Vol. 29, 1976.
18. *Annals of Internal Medicine,* Vol. 79, 1973.
19. June Biermann and Barbara Toohey, *The Diabetic's Total Health Book,* J. P. Tarcher, 1980.
20. *Utah Science,* Fall 1980.
21. *Diabetes Care,* December 1972; *American Journal of Clinical Nutrition,* May 1979, October 1980.
22. *Journal of Nutrition,* May 1977; *Practical Cardiology,* September 1980; *Archives of Internal Medicine,* Vol. 137, 1977.
23. *The Need for Special Foods and Sugar Substitutes by Individuals with Diabetes Mellitus,* Federation of American Societies for Experimental Biology, report to FDA, May 1978.
24. *Practical Cardiology,* September 1980, previously cited.
25. Fructose equaled sucrose in contributing to decay on the crown's smooth surfaces in human teeth, reported Theodore Koulourides, D.M.D., and associates from the Dental Institute,

University of Alabama, at the March 1974 meeting of the International Association for Dental Research.

26. *Contemporary Nutrition,* July 1980, August 1980.
27. The new method of producing fructose by using corn as a substrate is related, indirectly, to a process for the biological conversion of ethylene and propylene to their oxides. A halide ion and hydrogen peroxide are catalyzed with an enzyme, an alkene halohydrin. Then, a second enzyme converts the halohydrin to the oxide and releases the oxide for recycling. To make this process economical, it was necessary to find a ready supply of low-cost hydrogen peroxide. Another enzymatic process was developed to create hydrogen peroxide as a by-product, and therein is the fructose connection. The Cetus Corporation of Berkeley, California, a biological research company, screened thousands of enzymes before finding one capable of catalyzing glucose and oxygen to yield the desired hydrogen peroxide. In addition, it produced glucosone, which, in turn, can be catalyzed to fructose for food use. The new process is a joint venture of Cetus and Chevron (Standard Oil of California), a part owner of Cetus (*Business Week,* May 12, 1980; *Snack Food,* September 1980).

In 1980, Hoffmann–La Roche announced plans to produce crystalline fructose from corn syrup. First the syrup would be isomerized to HFCS, then the liquid fructose would be dried to crystalline form. The economics of making crystalline fructose from corn syrup is uncertain. The drying process is costly, and for feedstock the process competes against HFCS and gasohol. Fructose can be made from cornstarch, using a triple enzyme process. The final enzyme, isomerase, when added to dextrose, rearranges the dextrose atoms to form fructose. Both glucose and fructose have the same chemical composition, $C_6H_{12}O_6$. Isomerization reorients the glucose OH groups to form fructose (*Food Engineering,* May 1980).

In 1980, gene splicing was applied to HFCS processing. A bacterium capable of producing an enzyme that could convert corn syrup into 100 percent fructose was found; present methods in use allow only up to 90 percent conversion. Moreover, the new process would make it possible to produce fructose, from corn, in crystalline form. Efforts then were directed to locate and transfer the enzyme's gene from that bacterium into another bacterium. If these efforts are successful, enzyme fermentation could yield large-scale production. If perfected, such a process would have considerable impact on the food industry's favoring corn-based sweeteners rather than cane or beet sugars (*Food Processing*, November 1980).

28. *Wall Street Journal*, August 8, 1980.
29. *American Dairy Review*, June 1975.
30. *Food Processing*, June 1976.
31. *American Dairy Review*, June 1979.
32. *New York Times*, October 6, 1974.
33. *Dairy Field*, May 1980, June 1980.
34. *Food Processing*, July 1975, May 1976, August 1976, October 1976, May 1980; *Baking Industry*, September 1976, February 1977; *Dairy Record*, February 1980; March 1980; *Food Engineering*, April 1980; *Processed Prepared Foods*, April 1980.
35. *Business Week*, August 14, 1978.
36. *American Dairy Review*, June 1976.
37. *Food Engineering*, August 1979.
38. *American Agriculturist*, December 1978; *Forbes*, June 26, 1978.
39. *American Agriculturist*, February 1976.
40. *American Dairy Review*, June 1975, previously cited; *Dairy Record*, April 1979, May 1979.
41. *Baking Industry*, February 1977; *Dairy Record*, December 1979; *Dairy Field*, October 1979; *Food Engineering*, June 1978, August 1979; *Food Processing*, July 1977, October 1977, April 1978; *Food Product Development*, September 1979, December 1979; *Processed Prepared Foods*, August 1979.
42. *New York Times*, January 29, 1980; *Business Week*, August 14, 1978; *Food Product Development*, February 1980, November 1980.
43. *Nutrition Action*, February 1980.
44. *Federal Trade Commission News Summary*, July 11, 1980.
45. *Washington Post*, February 21, 1980.
46. *Food Engineering*, May 1980; *FDA Consumer*, March 1980.
47. *Processed Prepared Foods*, April 1980, previously cited.

48. *Chemical and Engineering News,* April 21, 1980.
49. *Chemistry,* Vol. 47, 1974; *Food Processing,* October 1974; *New York Times,* April 9, 1977; *Progressive Grocer,* June 1974; *Chemical and Engineering News,* May 1, 1973.
50. *Federal Register,* July 26, 1974; *New York Times,* July 26, 1974; *Food Product Development,* December 1974; *Chemical and Engineering News,* August 5, 1974; *Chemical Week,* August 7, 1974.
51. *Experimental Brain Research,* Vol. 14, 1971; *Brain Research,* Vol. 77, 1974; *Journal of Neuropathology and Experimental Neurology,* Vol. 34, 1975; *Sweeteners: Issues and Uncertainties,* Academy Forum, NAS, March 25/26, 1975, NAS/NRC, 1975; *Business Week,* August 10, 1974; *Medical World News,* June 7, 1974; *Chemical Week,* October 2, 1974.
52. *Food Processing,* May 1975; *Food Product Development,* February 1975.
53. *New England Journal of Medicine,* March 13, 1975; *Medical World News,* June 7, 1974; *Food Processing,* September 1974, October 1974.
54. *Federal Register,* September 24, 1974; *FDA Consumer,* February 1976.
55. *Food Chemical News,* November 5, 1973; *Food Processing,* August 1975.
56. *Chemical and Engineering News,* December 8, 1975; *New York Times,* November 17, 1976; *Wall Street Journal,* December 5, 1975; *Washington Post,* April 8, 1976; *Health, Education, and Welfare News,* April 8, 1976.
57. *Wall Street Journal,* April 9, 1976; *Regulation of the Food Additive Aspartame,* Report to the Comptroller General of the United States, April 8, 1976.
58. *Health, Education, and Welfare News,* May 31, 1979.
59. *Processed Prepared Foods,* March 1979.
60. *Health, Education, and Welfare News,* May 31, 1979, previously cited; *Food Product Development,* July 1979; *Chemical and Engineering News,* June 18, 1979; *Science News,* August 11, 1979.
61. *Food Product Development,* July 1979, previously cited, October 1979.
62. *Chemical and Engineering News,* August 13, 1979; *Business*

Week, September 29, 1980; *Food Product Development,* October 1979, November 1979.

63. *Science,* February 22, 1980; *Chemical and Engineering News,* February 18, 1980; *Food Engineering,* November 1980; *Food Product Development,* November 1980.

64. *Chemical and Engineering News,* October 6, 1980, July 20, 1981; *Science News,* July 25, 1981; *Newsweek,* July 27, 1981; *Health and Human Services News,* July 15, 1981.

65. *Food Engineering,* June 1981. Among tests conducted for Pfizer, one at Tulane University showed that maturity-onset diabetics could tolerate a 50-gram dose of polydextrose taken in a fasting state, with no significant blood glucose level increase or insulin demand, which suggested that polydextrose may be useful for diabetics. Another test, at the University of Zurich, suggested that polydextrose may be non-cariogenic.

66. *Science News,* May 2, 1981; *Food Processing,* July 1981; *Food Engineering,* August 1981.

Chapter 5. Artificial Sweeteners: Should Carcinogens Be Allowed? Cyclamates and saccharin

1. *Chemical and Engineering News,* April 6, 1976.

2. *Consumer Reports,* October 1964.

3. *Policy Statement on Artificial Sweeteners,* Food and Nutrition Board, National Academy of Science/National Research Council, adopted November 1954, published November 1955; *The Safety of Artificial Sweeteners for Use in Foods,* idem, Publication 386, August 1955.

4. *Fact,* November/December 1966.

5. *Chemical Abstracts,* Vol. 81, 1974, 24632.

6. *Policy Statement on Artificial Sweeteners,* revised, April 1962.

7. The greatly increased use of artificial sweeteners began in the 1960s. In 1960, the average U.S. annual consumption per person for saccharin was 1.9 pounds; cyclamates, 0.3 pounds (with saccharin being 300 to 500 times sweeter than sucrose; cyclamates 30 times sweeter). By 1968, cyclamate use peaked at 2.2 pounds, while saccharin use was 5.0 pounds. After the

cyclamates ban, saccharin rose steadily. By 1980, saccharin consumption rose to 7.1 pounds (*Sugar and Sweetener Report,* USDA, December 1976, February 1979, December 1980). These averages are figures that result from dividing the total population by total consumption. Since the average population includes segments who are non-users of artificial sweeteners, actual consumption by users is higher than the statistical figures.

8. *Regulation of Cyclamate Sweeteners.* Hearings, Committee on Government Operations, House of Representatives, October 10, 1970.

9. Philip Boffey, *The Brain Bank of America,* McGraw-Hill, 1975.

10. *The Medical Letter,* September 11, 1964.

11. *Connecticut Medicine,* August 1965.

12. *New York Times,* May 20, 1965.

13. *Journal of the American Dietetic Association,* June 1965; *Business Week,* March 22, 1969.

14. *Medical World News,* February 13, 1970.

15. *Nature,* March 25, 1967; *Journal of the American Medical Association,* September 4, 1967.

16. *Health Bulletin,* November 30, 1968.

17. *Science,* August 26, 1967.

18. *Capital Times* [Madison, Wis.], October 7, 1965; *Health Bulletin,* September 25, 1965; *Nature,* March 25, 1967, previously cited; *Proceedings of the Society for Experimental Biology and Medicine,* December 1969.

19. *Science News,* August 26, 1967.

20. *Washington Post,* November 17, 1968.

21. *Lancet,* December 13, 1969.

22. *Science,* September 12, 1969.

23. *Proceedings of the Society for Experimental Biology and Medicine.* December 1969, previously cited.

24. *Consumer Reports,* May 1969.

25. *Non-nutritive Sweeteners, Summary and Conclusions,* prepared by the Ad Hoc Committee on Non-nutritive Sweeteners, Food Protection Committee, NAS/NRC, November 1968.

26. *The Safety of Artificial Sweeteners,* Food and Drug Adminis-

tration Fact Sheet, April 3, 1969; *New York Times,* April 4, 1969.

27. *Business Week,* October 18, 1969.

28. *Medical Tribune,* July 6, 1970; *Archives of Environmental Health,* July 1971.

29. *Cyclamate Sweeteners,* Hearings, Committee on Government Operations, House of Representatives, June 10, 1970.

30. *Health Bulletin,* October 25, 1969.

31. *New York Times,* December 30, 1976.

32. *New York Times,* October 19, 1969; *U.S. News and World Report,* November 3, 1969.

33. *Barron's,* November 17, 1969.

34. *Consumer Reports,* January 1970.

35. *Medical Tribune,* August 31, 1970.

36. *New York Times,* November 23, 1969; *Food Chemical News,* November 24, 1969.

37. *Cyclamates,* Hearings, Committee on the Judiciary, House of Representatives, September 29 and 30, October 6, 1971.

38. *Nature,* January 4, 1969.

39. *New York Times,* August 15, 1970; *Evening News* [Newark, N.J.], September 10, 1970.

40. *Regulation of Cyclamate Sweeteners,* 36th Report by the Committee on Government Operations, U.S. Government Printing Office, October 8, 1970.

41. *Business Week,* March 25, 1972; *New Yorker,* February 4, 1974.

42. *Wall Street Journal,* August 7, 1970.

43. *Chemical and Engineering News,* September 27, 1971; *New York Times,* September 30, 1971, May 7, 1972; *Wall Street Journal,* March 10, 1972, July 25, 1972; *Business Week,* March 25, 1972.

44. *Food Chemical News,* August 2, 1980.

45. *Wall Street Journal,* July 2, 1973; *Chemical Week,* July 11, 1973; *Medical Tribune,* September 26, 1973, December 5, 1973; *Environment,* September 1973; *Medical World News,* March 15, 1974.

46. *Chemical and Engineering News,* November 26, 1973; *Business Week,* November 14, 1973.

47. *Medical World News,* March 15, 1974.
48. *Chemical and Engineering News,* September 16, 1974; *New York Times,* September 11, 1974; *Science,* November 1, 1974; *Medical World News,* October 11, 1974.
49. *Chemical and Engineering News,* October 21, 1974, November 18, 1974; *New York Times,* November 14, 1974.
50. *New York Times,* March 20, 1975; *Chemical and Engineering News,* March 24, 1975.
51. *Sweeteners: Issues and Uncertainties,* Academy Forum, NAS, March 25/26, 1975, NAS/NRC, 1975; *Chemical and Engineering News,* March 31, 1975, April 7, 1975.
52. *Star Ledger* [Newark, N.J.], December 11, 1975; *Wall Street Journal,* January 14, 1976; *International Medical News,* January 15, 1976; *Chemical and Engineering News,* March 15, 1976.
53. *FDA Consumer,* June 1976.
54. *Health, Education, and Welfare News,* May 11, 1976.
55. *Chemical and Engineering News,* June 21, 1976, October 9, 1978, October 23, 1978; *Wall Street Journal,* July 13, 1976; *Food Processing,* November 1976.
56. *Food Product Development,* November 1978, June 1979; *Health, Education, and Welfare News,* June 26, 1979.
57. *Chemical and Engineering News,* September 17, 1979, February 18, 1980; *Food Product Development,* October 1979, *Health and Human Services News,* September 4, 1980.
58. *Nature,* May 7, 1971.
59. *Chemical and Engineering News,* September 22, 1980; *Food Engineering,* October 1980; *Food Processing,* November 1980.
60. *Chemical and Engineering News,* April 6, 1976, April 11, 1977.
61. *Saccharin: Technical Assessments of Risks and Benefits,* Report No. 1, Committee for a Study on Saccharin and Food Safety Policy, NAS/NRC, November 1978.
62. *Good Housekeeping,* November 1913; Harvey W. Wiley, *An Autobiography,* 1930; Maurice Natenberg, *The Legacy of Dr. Wiley,* Regent House, 1957; Wiley, *The History of a Crime Against the Food Law,* 1929; Oscar E. Anderson, *The Health of a Nation,* 1958.

63. *British Medical Journal,* Vol. 1, 1897, Vol. 18, 1915; *Journal of the American Pharmaceutical Association,* August 1947; *New York State Journal of Medicine,* December 1, 1955. Among early reports, one experiment demonstrated that saccharin was a protoplasmic poison. Peas, first placed in a saccharin solution, failed to sprout (*Pharmaceutisch Week-blad,* Vol. 59, 1915). Saccharin was reported to induce cell proliferation and predisposed cells to cancer (*British Medical Journal,* October 9, 1915). One-cell animals died in solutions of one-part saccharin to 400-parts water. The stronger the saccharin solution, the less protein was digested. Saccharin inhibited the fat-digesting action of the pancreatic juices, and possibly had poisonous effects on cells in the bloodstream and in tissues. Saccharin was found to be 12 times more toxic to the one-cell animals than carbolic acid. The researchers concluded that extensive, prolonged human use of saccharin was unwise and objectionable and should be prohibited totally from the food supply, and restricted for drug use (*Archiv für Hygiene und Bakteriologie,* Vol. 92, 1923/1924). Saccharin's adverse chronic effects in mammals were reported (*Bollettino della Societa Italiana di Biologia Sperimentale,* Vol. 11, 1936). Saccharin was linked to goiter symptoms, both in human clinical cases and in animal experiments (*Bratislavski Lekarske Listy,* May 1938). Adverse symptoms from saccharin use, reported in various medical journals, included hives, itching, nausea, sweating, swelling and blistering of the tongue, fluttering in the ear, lowered blood sugar, irregular pulse and heart beat (*Journal of the American Medical Association,* November 1, 1965, September 4, 1967; *American Journal of Obstetrics and Gynecology,* September 15, 1971, August 15, 1972; *Cutis,* July 1972).

64. *Family Health,* October 1969.

65. *Policy Statement on Artificial Sweeteners,* November 1955, previously cited; *British Journal of Cancer,* Vol. 11, 1957.

66. *New York Times,* October 23, 1969.

67. *Wall Street Journal,* March 17, 1970.

68. *Science,* June 5, 1970; *Wall Street Journal,* March 18, 1970; *Archives of Environmental Health,* July 1971.

69. *New York Times,* March 20, 1970, July 23, 1970; *Washington Post,* July 23, 1970.

70. *Wall Street Journal,* September 18, 1970, June 23, 1971; *New York Times,* June 23, 1971.

71. *Chemical and Engineering News,* July 19, 1971.

72. *Health, Education, and Welfare News,* January 28, 1972.

73. *Business Week,* February 19, 1972; *Chemical Week,* February 16, 1972, March 25, 1972, April 1, 1972.

74. *Business Week,* November 27, 1978.

75. *Medical Tribune,* February 23, 1972.

76. *Health, Education, and Welfare News,* May 21, 1973.

77. *Nature,* June 8, 1973.

78. *Medical Tribune,* November 21, 1973.

79. *Wall Street Journal,* July 23, 1973.

80. *New York Times,* November 5, 1973.

81. *Safety of Saccharin and Sodium Saccharin in the Human Diet,* Subcommittee on Non-nutritive Sweeteners, Committee on Food Protection, Food and Nutrition Board, NAS/NRC, 1974.

82. *Health, Education, and Welfare News,* January 9, 1975.

83. *Need to Resolve Safety Questions on Saccharin,* Report of the Comptroller General of the United States, August 16, 1976.

84. *American Dairy Review,* February 1975; *Snack Food,* February 1975.

85. *New York Times,* October 20, 1976; *Star Ledger* [Newark, N.J.], January 7, 1977.

86. *Health, Education, and Welfare News,* March 9, 1977.

87. *New York Times,* March 11, 1977; March 12, 1977, March 13, 1977; *Newsweek,* March 21, 1977; *Science,* April 11, 1980.

88. *New York Times,* March 20, 1977.

89. *Capital Times* [Madison, Wis.], March 21, 1977.

90. *New York Times,* March 22, 1977.

91. James G. Wilson and Joseph Warkany, eds. *Teratology: Principles and Techniques,* University of Chicago Press, 1964.

92. *Science,* February 16, 1973.

93. *Newsday,* March 25, 1977; Samuel S. Epstein, M.D., *The Politics of Cancer,* Sierra Club, 1978.

94. *Journal of the American Dietetic Association,* Vol. 32, 1956;

Vol. 59, 1971, Vol. 63, 1973; *Proceedings of the Society for Experimental Biology and Medicine,* Vol. 66, 1947; *American Journal of Clinical Nutrition,* Vol. 7, 1959.

95. *Newsweek,* March 21, 1977, March 30, 1977.

96. *Science News,* March 19, 1977; *Washington Post,* May 3, 1977.

97. *Proposed Saccharin Ban,* Oversight Hearings, Health and Environment, Committee on Interstate Foreign Commerce, House of Representatives, March 21, 1977.

98. *Snack Food,* June 1977.

99. *Consumer Reports,* July 1977; *New York Times,* October 5, 1977; *Science,* April 11, 1980.

100. *Food Processing,* August 1977.

101. *Cancer Testing Technology and Saccharin,* Office of Technology Assessment, Congress of the United States, October 1977.

102. *Health, Education, and Welfare News,* April 14, 1977.

103. *New York Times,* April 5, 1977.

104. *New York Times,* May 19, 1977; *Chemical and Engineering News,* May 23, 1977.

105. *Chemical and Engineering News,* June 13, 1977.

106. *New York Times,* June 19, 1977; *Food and Drug Administration Talk Paper,* June 20, 1977; *Wall Street Journal,* June 20, 1977; *Chemical and Engineering News,* June 27, 1977.

107. *Chemical and Engineering News,* July 18, 1977; *Food Processing,* August 1977.

108. *Chemical and Engineering News,* September 26, 1977.

109. *Food Processing,* November 1977.

110. *Health, Education, and Welfare News,* November 28, 1977.

111. *Health, Education, and Welfare News,* January 25, 1978.

112. *FDA Consumer,* March 1980.

113. *New York Times,* November 7, 1978; *Chemical and Engineering News,* November 13, 1978; *Science,* September 22, 1978, November 24, 1978.

114. *New York Times,* March 3, 1979, March 5, 1979; *Science News,* March 10, 1979.

115. *New York Times,* May 10, 1979; *Chemical and Engineering News,* May 21, 1979; *Health, Education, and Welfare News,* May 22, 1979.

116. *Chicago Tribune,* May 4, 1979.

117. *Chemical and Engineering News,* May 28, 1979.

118. *Snack Food,* June 1979.

119. *Dairy and Ice Cream Field,* July 1979; *Chemical and Engineering News,* July 2, 1979, July 6, 1981.

120. *New York Times,* December 21, 1979; *Chemical and Engineering News,* January 7, 1980.

121. *Progress Report to FDA from NCI Concerning the National Bladder Cancer Study,* released December 1979; *Health, Education, and Welfare News,* December 20, 1979.

122. *Science,* March 14, 1980; *Chemical and Engineering News,* March 17, 1980.

123. *New England Journal of Medicine,* March 6, 1980.

124. *Science,* April 11, 1980.

Chapter 6. Rare Sugars: How Useful? How Safe? Sorbitol, mannitol, and xylitol

1. *Lancet,* December 25, 1971. While glycerol and propylene glycol are sugar polyols, they have not been discussed in the text since their food uses are other than as sweeteners.

2. Intravenously administered sorbitol is oxidized rapidly to fructose in the liver. The metabolic changes induced are similar to those induced by fructose (*Diabetes Care,* March/April 1980).

3. *Practical Cardiology,* September 1980; *Journal of the American Pharmaceutical Association,* Vol. 40, 1951.

4. William C. Griffin and Matthew J. Lynch, "Polyhydric Alcohols," in *Handbook of Food Additives,* Thomas E. Furia, ed., 1968.

5. *Evaluation of the Health Aspects of Sorbitol as a Food Ingredient,* prepared for the Food and Drug Administration by the Federation of American Societies for Experimental Biology, December 1972.

6. *Food Processing,* January 1977; *Canner/Packer '77 Buyers Guide; Baking Industry,* November 1976; *United States Department of Agriculture News,* November 8, 1972; *Health Food Retailing,* March 1973; *Snack Food,* December 1974.

7. *74th Report on Food Products,* 1969, Connecticut Agricultural Experiment Station, issued July 1970; *Consumer Affairs Newsletter* [Syracuse, N.Y.], February 1979.
8. *Diabetes Care,* March/April 1980.
9. *Contemporary Nutrition,* July 1980, August 1980.
10. Francis J. C. Roe, ed., *Metabolic Aspects of Food Safety,* Blackwell Scientific Publications, 1980.
11. *Journal of Biological Chemistry,* Vol. 141, 1941; *American Journal of Clinical Nutrition,* March/April 1960.
12. *Food,* USDA Yearbook of Agriculture, 1959. Sorbitol is a fairly common component as a vehicle in liquid pharmaceutical products, but little is known about its gastrointestinal effects. The same is true for glycerols and propylene glycol. These three vehicles were tested in adult rats and dogs, by means of stomach tube, three times daily for three days. When the substances were administered, undiluted, the mucosa of the stomach and duodenum showed irritation. Sorbitol was less irritating than glycerol, but more irritating than propylene glycol. The irritation of each substance decreased with increased dilution in distilled water (*Food and Cosmetics Toxicology,* December 1969).
13. *Diabetes Care,* March/April 1980, previously cited, July/August 1978.
14. *Gastroenterology,* Vol. 47, 1964; *Journal of Biological Chemistry,* Vol. 141, 1941; *The Need for Special Food and Sugar Substitutes by Individuals with Diabetes Mellitus,* Federation of American Societies for Experimental Biology, May 1978; *Lancet,* December 25, 1971, previously cited; *Journal of the American Dietetic Association,* November 1978.
15. *New England Journal of Medicine,* September 29, 1966.
16. *New England Journal of Medicine,* February 16, 1967.
17. *Journal of the American Medical Association,* July 18, 1980.
18. The side effects from high doses of most carbohydrates, including sorbitol, are well recognized by physicians. The effects include increased serum uric acid concentration; increased bilirubin; increased serum lactate level and lactate pyruvate ratios; decreased concentration of free fatty acids and phosphate; and a tendency toward acidosis. There is no significant side effect

after the oral administration of sorbitol at levels that are the same, calorically, as average amounts of table sugar added to foods. Rapid intravenous administration of sorbitol at high dosage may result in increased uric acid production and in a temporary decrease of hepatic inorganic phosphate (*Diabetes Care,* July/August 1978, previously cited).

19. The possible influence of sorbitol (and xylitol) on cataract formation was discussed by an Ad Hoc Study Group of FASEB, requested by FDA. The consensus was that exogenous sorbitol and xylitol have *no specific or direct* influence on cataract formation. This view is supported by the fact that the intact lens is nearly impermeable to sorbitol and xylitol. *Indirectly,* however, a possibility exists. Glucose, resulting from the metabolism of sorbitol and xylitol in the liver, could contribute to hyperglycemia in a poorly controlled diabetic. In this case, there is elevated glucose level in the aqueous humor and lens, which may result in conversion of the glucose via aldose reductase and reduced nicotinamide adenine dinucleotide phosphate to sorbitol in the lens. These events are considered as a major factor in diabetic cataract development (*The Need for Special Foods and Sugar Substitutes by Individuals with Diabetes Mellitus,* previously cited). Among ADA recommendations for future investigation were (1) more data on the metabolic effects of sorbitol in foods consumed in the regular diet of diabetics; (2) 24-hour and 30-day effects on blood sugar, insulin response, and other parameters as a result of administering various amounts of sorbitol and other sugars in different combinations in the diets of persons with mild adult-onset diabetes where fasting blood sugars are normal but glucose tolerance tests are abnormal; mild or moderately severe maturity-onset diabetes patients with persistent fasting hyperglycemia; and lean, juvenile-onset insulin-dependent diabetics. The effects of controlling blood glucose during the 24-hour period as well as from minute to minute are not well defined (*Diabetes Care,* July/August 1978, previously cited).

20. *Lancet,* May 26, 1973.
21. *Scientific American,* December 1975.
22. *Diabetes Care,* July/August 1978, previously cited.

23. *Federal Register,* June/August 1973, 121.90; *Food Chemical News,* November 5, 1973.

24. *Food Chemical News,* November 5, 1973, cited above.

25. *Contemporary Nutrition,* July 1980, August 1980, previously cited.

26. Experiments showed that flushing the bladder of mice with a form of mannitol prevented urinary tract infections by *Escherichia coli (Science News,* July 29, 1978). Mannitol reduces clot retention after prostatectomy (*Modern Medicine,* May 15, 1978). Since mannitol binds itself when compressed, it is used illicitly to cut heroin (*Chemical and Engineering News,* January 17, 1972).

27. *Food and Cosmetics Toxicology,* October 1971.

28. *New England Journal of Medicine,* July 12, 1979.

29. Xylose is used medicinally to test for cow's milk protein intolerance. One hour after being challenged with cow's milk, the blood xylose may drop by at least 50 percent in an intolerant individual (*Medical Tribune,* October 24, 1979).

30. *Chemical and Engineering News,* September 12, 1977, March 14, 1977, March 21, 1977.

31. *Snack Food,* July 1977.

32. *Science News,* December 10, 1977; *New York Times,* November 16, 1977; *Food Product Development,* February 1979; *Chemical and Engineering News,* November 6, 1978; *Let's Live,* August 1977; *Health Food Retailing,* July 1980.

33. Xylitol is not metabolized via the main metabolic process for sugar breakdown (glycolysis) but by an alternate metabolic process known as the "pentose shunt pathway." When additional amounts of xylitol are administered, transketolase, an enzyme in the pentose shunt pathway, becomes overloaded. Excessive xylitol may interfere with normal metabolism of ribose, an essential RNA building block. This increased RNA breakdown, and especially in its components, the purines, has been confirmed. Also, the breakdown product, uric acid, is found at elevated levels in human and animal blood after xylitol ingestion.

34. *New York Times,* November 16, 1977.

35. *Lancet,* December 25, 1971, previously cited. Clinical investi-

gations were conducted to learn whether xylitol infusions promote calcium oxalate crystallizations. Although none of the patients showed hyperoxalemia or hyperoxaluria after xylitol infusions, glycolate excretions increased two- or threefold, together with increased excretion of tetronic acids. It was suggested that while xylitol breakdown may generate oxalate precursors, oxalosis occurring in association with xylitol infusions is caused by some factor other than xylitol's metabolism. The most likely predisposing variable is abnormal kidney function.

36. *Science,* February 10, 1978.
37. *Food Processing,* April 1976; *American Journal of Clinical Nutrition,* March 1976.
38. *Journal of Nutrition,* March 1977.
39. *Science/Medical News Leads,* No. 5, University of Washington, November 15, 1979.
40. *New York Times,* April 24, 1978; *Let's Live,* August 1977; *Health Food Retailing,* July 1980.

Chapter 7. Future Sweeteners: How Promising? Licorice, miracle berry, serendipity berry, dihydrochalcones, katemfi, stevioside, osladin, amino acids, synthetic sweeteners, non-absorbable leached polymers, non-digestible sugar

1. *Food Engineering,* March 1979.
2. *Chemicals Used in Food Processing,* Publication No. 1274, NAS/NRC, 1965; *Baking Industry,* May 1976; *Canner/Packer,* September 1977; *Food Processing,* May 1977; *MD,* June 1979; *Food Engineering,* March 1979.
3. *Food Processing,* May 1977; *Processed Prepared Foods,* August 1978.
4. *Canner/Packer,* September 1977, previously cited; *Food Processing,* November 1977.
5. *Food Product Development,* July 1979, August 1979.
6. *Drug Therapy,* October 1978; *Toxicants Occurring Naturally in Foods,* NAS/NRC, July 1971; *Good Housekeeping,* November 1976.
7. *Journal of the American Medical Association,* August 24, 1970.

8. *Journal of the American Medical Association*, Vol. 205, 1968.

9. *Journal of the American Medical Association*, August 24, 1970, previously cited.

10. *Pennsylvania Medicine*, Vol. 74, 1971.

11. *New England Journal of Medicine*, August 21, 1980.

12. *New England Journal of Medicine*, April 10, 1980.

13. *Journal of the American Medical Association*, Vol. 205, 1968, previously cited.

14. *New York Times*, November 22, 1976.

15. *New York Times*, December 3, 1972, November 22, 1976, cited above; *Science*, December 8, 1972.

16. *Helvetica Chimica Acta*, Vol. 50, 1967; *Journal of Agricultural and Food Chemistry*, Vol. 17, 1969; *Drug and Cosmetic Industry*, August 1973.

17. *Science*, September 20, 1968; *Drug Therapy*, October 1978, previously cited; *New York Times*, June 10, 1971, November 22, 1976, previously cited; *Chemistry*, November 1971; *Food Chemical News*, October 1, 1973; *American Dairy Review*, January 1975. The active principle in the miracle berry is a glyco-protein — a protein with sugar groups attached — which coats the tongue in a very thin but durable layer. The glyco-protein contains 6.7 percent sugar (L-arabinose and D-xylose). The purified protein is tasteless. It is destroyed by boiling and by exposure to organic solvents (*Science*, Vol. 161, September 20, 1968). It is thought that the sweet sensation is produced by the protein's attaching itself to the receptors of the taste buds and modifying their function, as an anesthetic. This is analogous to the property of certain deodorizing preparations that function by anesthetizing the olfactory nerves temporarily and making it difficult to detect disagreeable odors (*American Dairy Review*, January 1975). Others do not believe that miraculin acts as an anesthetic, but rather that it influences the response to acid of the taste receptors on the palate in such a way as to change the recognition stimulation pattern from sour to sweet.

18. Press release, Ketchum, MacLeod, and Grove, Inc., November 15, 1972.

19. *Health, Education, and Welfare News,* May 23, 1977.
20. *Journal of the American Medical Association,* April 10, 1972; *Journal of Food Science,* Vol. 34, 1969. The large monellin molecule has 92 residues and an intact tertiary structure that is necessary to produce its sweetness. Monellin consists of two non-covalently bound chains of 50 and 42 residues. When separated, neither shows sweetness. But the two chains, recombined partially, form a tertiary structure that again shows sweetness (*Chemical and Engineering News,* August 25, 1975).
21. *New York Times,* April 9, 1977; *Chemical and Engineering News,* March 21, 1972; *Business Week,* March 23, 1972.
22. *MD,* June 1979, previously cited.
23. *USDA's Service,* May 1977; *Health Bulletin,* December 20, 1969; *Science News,* July 16, 1977; *United States Department of Agriculture Daily Summary,* March 18, 1977; *Food Processing/Foods of Tomorrow,* Summer 1970; *Food Processing,* May 1977; *Chemical and Engineering News,* August 25, 1975; *Processed Prepared Foods,* February 1979; *Journal of Agricultural and Food Chemistry,* Vol. 16, 1968; *New York Times,* March 16, 1977. One explanation of the slow onset and lingering aftertaste effect of DHC sweeteners may relate to the general property of phenols. They bind readily with proteins. DHC, with their phenolic groups, may bind rapidly and indiscriminately with protein in saliva. In that case, it would take time for the DHC activity to use up available saliva protein before binding to taste receptor sites. Three sites responsible for the sweetness have been located on DHC molecules and analogs. They fit a model in which the surface configuration of a sweetener has three sites at specific angles to each other. This idea is given support by another observation. The DHC molecules have two phenol rings, with part of the three activating sites distributed between them. Saccharin and cyclamates have only one similar site per molecule. It is possible that both of these compounds must pair up with a single receptor on the tongue to create the sweetness sensation.
24. *Food Processing,* May 1977; *MD,* June 1979, previously cited.
25. *New York Times,* April 9, 1977, previously cited; *Chemical and Engineering News,* March 27, 1972; *Sweeteners: Issues*

and Uncertainties, previously cited; *European Journal of Biochemistry,* Vol. 31, 1972; Vol. 35, 1973.

26. *American Perfumes and Essential Oil Review,* Vol. 66, 1955.

27. *Drug Therapy,* October 1978, previously cited.

28. *Sweeteners: Issues and Uncertainties,* previously cited. Osladin, derived from *Polypodium vulgare,* is a steroidal saponin and the only one that is not bitter (*Drug Therapy,* October 1978, cited above).

29. Vernal S. Packard, Jr., *Processed Foods and the Consumer,* University of Minnesota Press, 1976; *Chemistry and the Food System,* NAS/NRC, 1980; *Conference on Foods, Nutrition, and Dental Health,* American Dental Association Health Foundation, October 1978; *Science News,* May 30, 1970.

30. *Processed Prepared Foods,* February 1979; *Food Processing,* May 1977.

31. *Newsweek,* April 4, 1977; *Chemical and Engineering News,* August 25, 1975. SRI oxime V is oxathiazinone dioxide, derived from several sources and is not yet defined. Perillartin is used in Japan as a tobacco-sweetening substitute for licorice, maple sugar, or honey.

32. *Chemical and Engineering News,* August 25, 1975, cited above; *Business Week,* May 14, 1979; *Food Processing,* January 1979; *Food Engineering,* September 1978; *Food Product Development,* October 1978.

33. A current approach to make sugar substitutes is by use of sugar itself. By attaching chlorine atoms onto sucrose molecules, sweet compounds can be created and considered noncaloric since they are not digested by the body. Building on sucrose, chemists have made one compound that is 2000 times sweeter than sugar, another that is bitter, and one that is tasteless ("Sweet Solutions," *Nova* television program, National Science Foundation, 1979).

34. *Science News,* May 2, 1981; *Food Engineering,* July 1981.

Chapter 8. The Sugar Trap: How Can We Avoid It?

1. *Co-Op News* [Hanover, N.H.], March 1975.

2. *New York Times,* August 22, 1974.

3. *Business Week*, June 1, 1981.

4. Dr. Abraham Nizel, quoted in *Edible TV, Your Child and Food Commercials,* prepared by the Council on Children, Media, and Merchandising, for the Select Committee on Nutrition and Human Needs, United States Senate, September 1977; *American Journal of Clinical Nutrition,* February 1980.

5. *Nutrition Notes,* December 1970; *Snack Food,* May 1978; *Wall Street Journal,* June 7 ,1978.

6. United States Department of Agriculture press release, April 2, 1978; *Nation's Restaurant News,* May 29, 1978.

7. *Snack Food,* June 1978.

8. *Chemical and Engineering News,* April 7, 1975; *Progressive Grocer,* July 1975; *Food Product Development,* November 1979; *Processed Prepared Foods,* July 1978; *Food Engineering,* March 1979.

9. *Co-Op News* [Hanover, N.H.], September 1977; *FDA Consumer,* April 1980.

10. *Nutrition and Your Health, Dietary Guidelines for Americans,* Washington, D.C., United States Department of Agriculture/ Health, Education, and Welfare, February 1980.

Selected Bibliography

Books and articles on sugars:

Ahrens, Richard A. "Sucrose, Hypertension, and Heart Disease: An Historical Perspective." *American Journal of Clinical Nutrition,* April 1974.

Anderson, Oscar E. *Health of a Nation: Harvey W. Wiley's fight for pure food.* Chicago: University of Chicago, 1958.

Biermann, June, and Barbara Toohey. *The Diabetic's Total Health Book.* Los Angeles: J.P. Tarcher, 1980.

Boffey, Philip. *The Brain Bank of America: An Inquiry into the Politics of Science.* New York: McGraw-Hill, 1975.

Cleave, T. L., Surgeon-Captain, Royal Navy. *The Saccharine Disease.* New Canaan, Conn.: Keats Publishing, 1975.

Conner, William E., and Sonja L. Conner. "Sucrose and Carbohydrates," in *Present Knowledge in Nutrition,* 4th ed. Washington, D. C.: The Nutrition Foundation, Inc., 1976.

Friend, Bertha. "Nutrients in the U.S. Food Supply: A Review of Trends, 1909–1913 to 1965." *American Journal of Clinical Nutrition,* Vol. 20, 1967.

Furia, Thomas E. *Handbook of Food Additives.* Cleveland: The Chemical Rubber Co., 1968.

Hess, John L. "Harvard's Sugar-Pushing Nutritionist." *Saturday Review,* August 1978.

Mayer, Jean. "The Bitter Truth About Sugar." *New York Sunday Times Magazine,* June 20, 1976.

Newbrun, Ernest, "Dietary Factors in Dental Decay." *Nutrition and the M.D.,* October 1976.

Randolph, Theron G., M.D. "Allergic Reactions from the Ingestion or Intravenous Injection of Cane Sugar (Sucrose)." *Journal of Laboratory Clinical Medicine,* Vol. 36, 1950.

————. "Beet Sensitivity: Allergic Reactions from the Ingestion of Beet Sugar (Sucrose) and Monosodium Glutamate of Beet Origin." *Journal of Laboratory Clinical Medicine,* Vol. 36, 1950.

————. "Corn Sugar as an Allergen." *Annals of Allergy,* September/October 1949.

————. "The Role of Specific Sugars" in *Clinical Ecology.* Lawrence D. Dickey, M.D., ed. Springfield, Ill.: Charles C. Thomas, 1976.

————, and Ralph W. Moss, Ph.D. *An Alternative Approach to Allergies.* New York: Lippincott and Crowell, 1980.

Saccharin: Technical Assessment of Risks and Benefits. Report No. 1, Committee for a Study on Saccharin and Food Safety Policy. Washington, D.C.: Assembly of Life Sciences/Institute of Medicine, NAS/NRC, November 1978.

Safety of Saccharin and Sodium Saccharin in the Human Diet. Subcommittee on Non-nutritive Sweeteners, Committee on Food Protection, Food and Nutrition Board, NAS/NRC. Washington, D.C., 1974.

Shannon, Ira L., D.M.D., M.S.D. *Brand Name Guide to Sugar.* Chicago: Nelson-Hall, 1977.

Sipple, Horace L., and Kristen W. McNutt, eds. *Sugars in Nutrition.* New York: Academic Press, 1974.

"Sugar: Dangerous to the Heart?" *Medical World News,* February 12, 1971.

Sugars in Nutrition, a Brief Discussion. [leaflet], Washington, D.C.: United Fresh Fruit and Vegetable Association, October 1971.

Sweeteners: Issues and Uncertainties. Academy Forum. Washington, D.C.: National Academy of Sciences, 1975.

Wiley, Harvey W., M.D. *An Autobiography.* New York: Bobbs Merrill, 1930.

————. *The History of a Crime Against the Food Law.* Washington, D.C.: Harvey W. Wiley, 1929.

Yudkin, John M.D. "Sucrose and Heart Disease." *Nutrition Today,* Spring 1969.

————. *Sweet and Dangerous.* New York: Peter H. Wyden, 1972.

Government publications (request prices from Superintendent of Documents, U.S. Government Printing Office, Washington, D.C. 20402):

Dietary Goals for the United States. Select Committee on Nutrition and Human Needs, U.S. Senate, rev. 2nd. ed., February 1977.

Edible TV: Your Child and Food Commercials, prepared by the Council on Children, Media, and Merchandising, for the Select Committee on Nutrition and Human Needs, U.S. Senate, September 1977.

Food Safety: Where Are We? prepared for the Committee on Agriculture, Nutrition, and Forestry. U.S. Senate, July 1979.

Regulation of Cyclamate Sweeteners. 36th Report, Committee on Government Operations, House of Representatives, Report 91-1585, 1970.

The Hassle-Free Guide to a Better Diet. Leaflet No. 567. USDA Science and Education Administration, March 1980.

What's to Eat? And Other Questions Kids Ask About Food. USDA Yearbook, 1979.

Government reports requested by FDA: evaluation of GRAS list ingredients, prepared by the Federation of American Societies for Experimental Biology (FASEB) for the Food and Drug Administration (available from U.S. Department of Commerce, National Technical Information Service, Springfield, Virginia 22161):

Corn Sugar (Dextrose), Corn Syrup and Invert Sugar. Document No. PB-262 659, $6.50.

Dextrins, Document No. PB-254 538, $5.

Evaluation of GRAS Monographs. Document No. PB-80-203 789, $6.50.

Glycyrrhiza. Document No. PB-254 529, $6.50.

Mannitol. Document No. PB-221 953, $5.

Sorbitol. Document No. PB-221 951, $3.

Sorbose, Document No. PB-254 525, $5.

Sucrose. Document No. PB-262 668, $6.50.

Also available from the National Technical Information Service:

Safety of Saccharin and Sodium Saccharin in the Human Diet. Subcommittee on Nonnutritive Sweeteners, Committee on Food Protection, Food and Nutrition Board, National Research Council, 1974, $4.75.

Government hearings (request prices from Superintendent of Documents):

Banning of Saccharin, 1977. Subcommittee on Health and Scientific Research, Committee on Human Resources, U.S. Senate, June 1, 1977.

Cyclamate Sweeteners. Committee on Government Operations, House of Representatives, June 10, 1970.

Cyclamates. Committee on the Judiciary, House of Representatives, September 29, 30, and October 6, 1971.

Hyperactive Children. Subcommittee on Health, Committee on Labor and Public Welfare/Subcommittee on Administrative Practice and Procedure, Committee on the Judiciary, U.S. Senate, September 11, 1975.

Nutritional Content and Advertising for Dry Breakfast Cereals. Consumer Subcommittee, Committee on Commerce, U.S. Senate, July 23, August 4, 5, 1970; March 2, 1972.

Proposed Saccharin Ban. Oversight. Subcommittee on Health and the Environment, Committee on Interstate and Foreign Commerce, House of Representatives, March 21, 1977.

Sugar in Diet, Diabetes, and Heart Disease. Select Committee on Nutrition and Human Needs, U.S. Senate, April 30 and May 1, 2, 1973.

Government reports available from federal agencies:

FEDERAL TRADE COMMISSION:
Federal Trade Commission Staff Report on Television Advertising to Children. Washington, D.C. 20580: Federal Trade Commission, February 1978.

234   SELECTED BIBLIOGRAPHY

Proposed Trade Regulation Rule: Food Advertising. Report of the
    Presiding President. Public Record 215-40. Washington, D.C.
    20580: Federal Trade Commission, February 21, 1978.
*Proposed Trade Regulation Rule on Food Advertising. Staff Report
    and Recommendations.* Washington, D.C. 20580: Federal
    Trade Commission, September 25, 1978.

GENERAL ACCOUNTING OFFICE:
Need to Resolve Safety Questions on Saccharin. Report of the
    Comptroller General of the United States. Washington, D.C.
    20548: General Accounting Office, August 16, 1976.
Better Information Needed on Nutritional Quality of Foods. Report
    to the Congress of the United States by the Comptroller Gen-
    eral. Washington, D.C. 20548: General Accounting Office,
    April 30, 1980.

U.S. DEPARTMENT OF AGRICULTURE:
Nutrition and Your Health: Dietary Guidelines for Americans.
    Washington, D.C. 20250: USDA/HEW, February 1980.
Sugar and Sweetener Report (a monthly publication, mainly statis-
    tical charts showing sugar sales, uses, consumption, etc.)
    Washington, D.C. 20250: United States Department of Agri-
    culture.

U.S. DEPARTMENT OF HEALTH AND HUMAN SERVICES:
*Progress Report to the Food and Drug Administration from the
    National Cancer Institute Concerning the National Bladder
    Cancer Study.* Bethesda, Md. 20205: Office of Cancer Com-
    munications, U.S. Public Health Service, National Cancer In-
    stitute, 1979.
*The Need for Special Food and Sugar Substitutes by Individuals
    with Diabetes Mellitus.* Washington, D.C. 20204: Bureau of
    Foods, Food and Drug Administration, May 1978.

*Special materials for children, concerned parents, and groups
interested in upgrading school food (request prices from ad-
dresses):*

Burns, Marilyn. *Good for Me! All About Food in 32 Bites* (written
    for children). Boston: Little, Brown, 1978.

Goodwin, Mary T., Ph.D., and Gerry Pollen. *Creative Food Experiences for Children* (written by a public health nutritionist and early childhood elementary school teacher). Center for Science in the Public Interest, 1755 S Street NW, Washington, D.C. 20009, revised in 1980.

Kinsella, Susan. *Food on Campus* (a step-by-step guide to improve college food service). Emmaus, Pa. 18049: Rodale Press, 1978.

Lansky, Vicki. *Feed Me! I'm Yours* (a recipe book for mothers of young children, which attempts to reduce sugars in the diet). New York: Bantam, 1979.

———. *The Taming of the C.A.N.D.Y.* Monster (*Continuously Advertised Nutritionally Deficient Yummies)*. 1978. Meadowbrook Press, 166–48 Meadowbrook Lane, Wayzata, Minn. 55391.

Lisciandro, Frank J. *The Sugar Film* (film available for rental or purchase). 1980. Images Associates, 352 Conejo Rd., Santa Barbara, Calif. 93103.

Moyer, Anne. *Better Food for Public Places: A Guide for Improving Institutional Food*. Emmaus, Pa. 18049: Rodale Press, 1977.

Sloan, Sara. *Children Cook Naturally*. 1980.

———. *From Classroom to Cafeteria: A Nutritional Guide for Teachers and Managers*. Revised April 1978.

———. *A Guide for Nutra Lunches and Natural Foods*. 1977.

———. *Yuk to Yum Snacks*. 1978.

Sara Sloan Nutra Program, Box 13825, Atlanta, Ga. 30324. (Sloan is director of the Food Service Program of Fulton County Schools, Atlanta, Ga., and has been instrumental in upgrading school food.)

Turner, Mary and James. *Making Your Own Baby Food*. New York: Bantam, 1973.

Wallace, James F. and Maureen J. *The Effects of Excessive Consumption of Refined Sugar on Learning Skills, Behavior, Attitudes and/or Physical Condition in School-Aged Children*. 1978. Parents for Better Nutrition, 33 N. Central St., Medford, Or. 97501.

Index